Richard P. Feynman was o[...] theoretical physicists and [...] Rockaway, New York, in 1918 [...] [M]assachusetts Institute of Technology, where [h]e graduated with a BS in 1939. He went on to Princeton and received his PhD in 1942. During the war years he worked at the Los Alamos Scientific Laboratory. He became Professor of Theoretical Physics at Cornell University, where he worked with Hans Bethe. He basically rebuilt the theory of quantum electrodynamics and for this work shared the Nobel Prize in 1965. His simplified rules of calculation became standard tools of theoretical analysis in both quantum electrodynamics and high-energy physics. Feynman was a visiting professor at the California Institute of Technology in 1950 and later that year accepted a permanent faculty appointment. He became Richard Chace Tolman Professor of Theoretical Physics in 1959. Over the years he worked with Nobel laureate Murray Gell-Mann on a theory for weak interactions; he formulated a mathematical theory that explained a whole range of properties of liquid helium at very low temperatures; and he did theoretical work on how the structure of the proton is revealed in bombardment by high-energy electrons. He had an extraordinary ability to communicate his science to audiences at all levels and was a well-known and popular lecturer. Series of his lectures were collected and published; these include *The Feynman Lectures on Physics* and *The Character of Physical Law* as well as this volume. His memoirs, *Surely You're Joking, Mr. Feynman*, were published in 1985 and became a surprising bestseller. He died in 1988 after a long illness. Freeman Dyson, of the Institute for Advanced Study in Princeton, New Jersey, called Richard Feynman 'the most original mind of his generation', while *The New York Times* in its obituary described him as 'arguably the most brilliant, iconoclastic and influential of the postwar generation of theoretical physicists'.

QED

THE STRANGE THEORY OF LIGHT AND MATTER

RICHARD P. FEYNMAN

PENGUIN BOOKS

PENGUIN BOOKS

Published by the Penguin Group
Penguin Books Ltd, 27 Wrights Lane, London W8 5TZ, England
Penguin Books USA Inc., 375 Hudson Street, New York, New York 10014, USA
Penguin Books Australia Ltd, Ringwood, Victoria, Australia
Penguin Books Canada Ltd, 10 Alcorn Avenue, Toronto, Ontario, Canada M4V 3B2
Penguin Books (NZ) Ltd, 182–190 Wairau Road, Auckland 10, New Zealand

Penguin Books Ltd, Registered Offices: Harmondsworth, Middlesex, England

First published in the USA by Princeton University Press 1985
Published in Great Britain by Penguin Books 1990
9 10 8

Printed in England by Clays Ltd, St Ives plc
Filmset in Linotron Baskerville

Contents

Foreword

The Alix G. Mautner Memorial Lectures were conceived in honor of my wife Alix, who died in 1982. Although her career was in English literature, Alix had a long and abiding interest in many scientific fields. Thus it seemed fitting to create a fund in her name that would support an annual lecture series with the objective of communicating to an intelligent and interested public the spirit and achievements of science.

I am delighted that Richard Feynman has agreed to give the first series of lectures. Our friendship goes back fifty-five years to our childhood in Far Rockaway, New York. Richard knew Alix for about twenty-two years, and she long sought to have him develop an explanation of the physics of small particles that would be understandable to her and to other non-physicists.

As an added note, I would like to express my appreciation to those who contributed to the Alix G. Mautner Fund and thus helped make these lectures possible.

LEONARD MAUTNER
Los Angeles, California
May 1983

Preface

Richard Feynman is legendary in the world of physics for the way he looks at the world: taking nothing for granted and always thinking things out for himself, he often attains a new and profound understanding of nature's behavior—with a refreshing and elegantly simple way to describe it.

He is also known for his enthusiasm in explaining physics to students. After turning down countless offers to give speeches at prestigious societies and organizations, Feynman is a sucker for the student who comes by his office and asks him to talk to the local high school physics club.

This book is a venture that, as far as we know, has never been tried. It is a straightforward, honest explanation of a rather difficult subject—the theory of quantum electrodynamics—for a nontechnical audience. It is designed to give the interested reader an appreciation for the kind of thinking that physicists have resorted to in order to explain how Nature behaves.

If you are planning to study physics (or are already doing so), there is nothing in this book that has to be "unlearned": it is a complete description, accurate in every detail, of a framework onto which more advanced concepts can be attached without modification. For those of you who have already studied physics, it is a revelation of what you were *really* doing when you were making all those complicated calculations!

As a boy, Richard Feynman was inspired to study calculus from a book that began, "What one fool can do, another can." He would like to dedicate this book to his readers with similar words: "What one fool can understand, another can."

RALPH LEIGHTON
Pasadena, California
February 1985

Acknowledgment

This book purports to be a record of the lectures on quantum electrodynamics I gave at UCLA, transcribed and edited by my good friend Ralph Leighton. Actually, the manuscript has undergone considerable modification. Mr. Leighton's experience in teaching and in writing was of considerable value in this attempt at presenting this central part of physics to a wider audience.

Many "popular" expositions of science achieve apparent simplicity only by describing something different, something considerably distorted from what they claim to be describing. Respect for our subject did not permit us to do this. Through many hours of discussion, we have tried to achieve maximum clarity and simplicity without compromise by distortion of the truth.

QED

1

Introduction

Alix Mautner was very curious about physics and often asked me to explain things to her. I would do all right, just as I do with a group of students at Caltech that come to me for an hour on Thursdays, but eventually I'd fail at what is to me the most interesting part: We would always get hung up on the crazy ideas of quantum mechanics. I told her I couldn't explain these ideas in an hour or an evening—it would take a long time—but I promised her that someday I'd prepare a set of lectures on the subject.

I prepared some lectures, and I went to New Zealand to try them out—because New Zealand is far enough away that if they weren't successful, it would be all right! Well, the people in New Zealand thought they were okay, so I guess they're okay—at least for New Zealand! So here are the lectures I really prepared for Alix, but unfortunately I can't tell them to her directly, now.

What I'd like to talk about is a part of physics that is *known*, rather than a part that is unknown. People are always asking for the latest developments in the unification of this theory with that theory, and they don't give us a chance to tell them anything about one of the theories that we know pretty well. They always want to know things that we don't know. So, rather than confound you with a lot of half-cooked, partially analyzed theories, I would like to tell you about a subject that has been very thoroughly analyzed.

I love this area of physics and I think it's wonderful: it is called quantum electrodynamics, or QED for short.

My main purpose in these lectures is to describe as accurately as I can the strange theory of light and matter—or more specifically, the interaction of light and electrons. It's going to take a long time to explain all the things I want to. However, there are four lectures, so I'm going to take my time, and we will get everything all right.

Physics has a history of synthesizing many phenomena into a few theories. For instance, in the early days there were phenomena of motion and phenomena of heat; there were phenomena of sound, of light, and of gravity. But it was soon discovered, after Sir Isaac Newton explained the laws of motion, that some of these apparently different things were aspects of the same thing. For example, the phenomena of sound could be completely understood as the motion of atoms in the air. So sound was no longer considered something in addition to motion. It was also discovered that heat phenomena are easily understandable from the laws of motion. In this way, great globs of physics theory were synthesized into a simplified theory. The theory of gravitation, on the other hand, was not understandable from the laws of motion, and even today it stands isolated from the other theories. Gravitation is, so far, not understandable in terms of other phenomena.

After the synthesis of the phenomena of motion, sound, and heat, there was the discovery of a number of phenomena that we call electrical and magnetic. In 1873 these phenomena were synthesized with the phenomena of light and optics into a single theory by James Clerk Maxwell, who proposed that light is an electromagnetic wave. So at that stage, there were the laws of motion, the laws of electricity and magnetism, and the laws of gravity.

Around 1900 a theory was developed to explain what matter was. It was called the electron theory of matter, and it said that there were little charged particles inside of atoms. This theory evolved gradually to include a heavy nucleus with electrons going around it.

Attempts to understand the motion of the electrons going around the nucleus by using mechanical laws—analogous to the way Newton used the laws of motion to figure out how the earth went around the sun—were a real failure: all kinds of predictions came out wrong. (Incidentally, the theory of relativity, which you all understand to be a great revolution in physics, was also developed at about that time. But compared to this discovery that Newton's laws of motion were quite wrong in atoms, the theory of relativity was only a minor modification.) Working out another system to replace Newton's laws took a long time because phenomena at the atomic level were quite strange. One had to lose one's common sense in order to perceive what was happening at the atomic level. Finally, in 1926, an "uncommon-sensy" theory was developed to explain the "new type of behavior" of electrons in matter. It looked cockeyed, but in reality it was not: it was called the theory of quantum mechanics. The word "quantum" refers to this peculiar aspect of nature that goes against common sense. It is this aspect that I am going to tell you about.

The theory of quantum mechanics also explained all kinds of details, such as why an oxygen atom combines with two hydrogen atoms to make water, and so on. Quantum mechanics thus supplied the theory behind chemistry. So, fundamental theoretical chemistry is really physics.

Because the theory of quantum mechanics could explain all of chemistry and the various properties of substances, it was a tremendous success. But still there was the problem of the interaction of light and matter. That is, Maxwell's theory of electricity and magnetism had to be changed to be in accord with the new principles of quantum mechanics that had been developed. So a new theory, the quantum theory of the interaction of light and matter, which is called by the horrible name "quantum electrodynamics," was finally developed by a number of physicists in 1929.

But the theory was troubled. If you calculated something roughly, it would give a reasonable answer. But if you tried to compute it more accurately, you would find that the correction you thought was going to be small (the next term in a series, for example) was in fact very large—in fact, it was *infinity*! So it turned out you couldn't really compute *anything* beyond a certain accuracy.

By the way, what I have just outlined is what I call a "physicist's history of physics," which is never correct. What I am telling you is a sort of conventionalized myth-story that the physicists tell to their students, and those students tell to their students, and is not necessarily related to the actual historical development, which I do not really know!

At any rate, to continue with this "history," Paul Dirac, using the theory of relativity, made a relativistic theory of the electron that did not completely take into account all the effects of the electron's interaction with light. Dirac's theory said that an electron had a magnetic moment—something like the force of a little magnet—that had a strength of exactly 1 in certain units. Then in about 1948 it was discovered in experiments that the actual number was closer to 1.00118 (with an uncertainty of about 3 on the last digit). It was known, of course, that electrons interact with light, so some small correction was expected. It was also expected that this correction would be understandable from the new theory of quantum electrodynamics. But when it was calculated, instead of 1.00118 the result was infinity—which is wrong, experimentally!

Well, this problem of how to calculate things in quantum electrodynamics was straightened out by Julian Schwinger, Sin-Itiro Tomonaga, and myself in about 1948. Schwinger was the first to calculate this correction using a new "shell game"; his theoretical value was around 1.00116, which was close enough to the experimental number to show that we were on the right track. At last, we had a quantum theory of electricity and magnetism with which we could calculate! This is the theory that I am going to describe to you.

The theory of quantum electrodynamics has now lasted for more than fifty years, and has been tested more and more accurately over a wider and wider range of conditions. At the present time I can proudly say that there is *no significant difference* between experiment and theory!

Just to give you an idea of how the theory has been put through the wringer, I'll give you some recent numbers: experiments have Dirac's number at 1.00115965221 (with an uncertainty of about 4 in the last digit); the theory puts it at 1.00115965246 (with an uncertainty of about five times as much). To give you a feeling for the accuracy of these numbers, it comes out something like this: If you were to measure the distance from Los Angeles to New York to this accuracy, it would be exact to the thickness of a human hair. That's how delicately quantum electrodynamics has, in the past fifty years, been checked—both theoretically and experimentally. By the way, I have chosen only one number to show you. There are other things in quantum electrodynamics that have been measured with comparable accuracy, which also agree very well. Things have been checked at distance scales that range from one hundred times the size of the earth down to one-hundredth the size of an atomic nucleus. These numbers are meant to intimidate you into believing that the theory is probably not too far off! Before we're through, I'll describe how these calculations are made.

I would like to again impress you with the vast range of phenomena that the theory of quantum electrodynamics describes: It's easier to say it backwards: the theory describes *all* the phenomena of the physical world except the gravitational effect, the thing that holds you in your seats (actually, that's a combination of gravity and politeness, I think), and radioactive phenomena, which involve nuclei shifting in their energy levels. So if we leave out gravity and radioactivity (more properly, nuclear physics), what have we got left? Gasoline burning in automobiles, foam and bubbles, the hardness of salt or copper, the stiffness of steel. In fact, biologists are trying to interpret as much as they can about life in terms of chemistry, and as I already explained, the theory behind chemistry is quantum electrodynamics.

I must clarify something: When I say that all the phenomena of the physical world can be explained by this theory, we don't really know that. Most phenomena we are familiar with involve such *tremendous* numbers of electrons that it's hard for our poor minds to follow that complexity. In such situations, we can use the theory to figure roughly what ought to happen and that *is* what happens, roughly, in those circumstances. But if we arrange in the laboratory an experiment involving just a *few* electrons in *simple* circumstances, then we can calculate what might happen very accurately, and we can measure it very accurately, too. Whenever we do such experiments, the theory of quantum electrodynamics works very well.

We physicists are always checking to see if there is something the matter with the theory. That's the game, because if there *is* something the matter, it's interesting! But so far, we have found nothing wrong with the theory of quantum electrodynamics. It is, therefore, I would say, the jewel of physics—our proudest possession.

The theory of quantum electrodynamics is also the prototype for new theories that attempt to explain nuclear phenomena, the things that go on inside the nuclei of atoms. If one were to think of the physical world as a stage, then the actors would be not only electrons, which are outside the nucleus in atoms, but also quarks and gluons and so forth—dozens of kinds of particles—inside the nucleus. And though these "actors" appear quite different from one another, they all act in a certain style—a strange and peculiar style—the "quantum" style. At the end, I'll tell you a little bit about the nuclear particles. In the meantime, I'm only going to tell you about photons—particles of light—and electrons, to keep it simple. Because it's the *way* they act that is important, and the way they act is very interesting.

So now you know what I'm going to talk about. The next question is, will you *understand* what I'm going to tell you? Everybody who comes to a scientific lecture knows they are not going to understand it, but maybe the lecturer has a nice, colored tie to look at. Not in this case! (Feynman is not wearing a tie.)

What I am going to tell you about is what we teach our physics students in the third or fourth year of graduate school—and you think I'm going to explain it to you so you can understand it? No, you're not going to be able to understand it. Why, then, am I going to bother you with all this? Why are you going to sit here all this time, when you won't be able to understand what I am going to say? It is my task to convince you *not* to turn away because you don't understand it. You see, my physics students don't understand it either. That is because *I* don't understand it. Nobody does.

I'd like to talk a little bit about understanding. When we have a lecture, there are many reasons why you might not understand the speaker. One is, his language is bad—he doesn't say what he means to say, or he says it upside down—and it's hard to understand. That's a rather trivial matter, and I'll try my best to avoid too much of my New York accent.

Another possibility, especially if the lecturer is a physicist, is that he uses ordinary words in a funny way. Physicists often use ordinary words such as "work" or "action" or "energy" or even, as you shall see, "light" for some technical purpose. Thus, when I talk about "work" in physics, I don't mean the same thing as when I talk about "work" on the street. During this lecture I might use one of those words without noticing that it is being used in this unusual way. I'll try my best to catch myself—that's my job—but it is an error that is easy to make.

The next reason that you might think you do not understand what I am telling you is, while I am describing to you *how* Nature works, you won't understand *why* Nature works that way. But you see, nobody understands that. I can't explain why Nature behaves in this peculiar way.

Finally, there is this possibility: after I tell you something, you just can't believe it. You can't accept it. You don't like it. A little screen comes down and you don't listen anymore. I'm going to describe to you how Nature is—and if you don't like it, that's going to get in the way of your understanding it. It's a problem that physicists have learned to deal with: They've learned to realize that whether they like a theory or they don't like a theory is *not* the essential question. Rather, it is whether or not the theory gives predictions that agree with experiment. It is not a question of whether a theory is philosophically delightful, or easy to understand, or perfectly reasonable from the point of view of common sense. The theory of quantum electrodynamics describes Nature as absurd from the point of view of common sense. And it agrees fully with experiment. So I hope you can accept Nature as She is—absurd.

I'm going to have fun telling you about this absurdity, because I find it delightful. Please don't turn yourself off because you can't believe Nature is so strange. Just hear me all out, and I hope you'll be as delighted as I am when we're through.

How am I going to explain to you the things I don't explain to my students until they are third-year graduate students? Let me explain it by analogy. The Maya Indians were interested in the rising and setting of Venus as a morning "star" and as an evening "star"—they were very interested in when it would appear. After some years of observation, they noted that five cycles of Venus were very nearly equal to eight of their "nominal years" of 365 days (they were aware that the true year of seasons was different and they made calculations of that also). To make calculations, the Maya had invented a system of bars and dots to represent numbers (including zero), and had rules by which to calculate and predict not only the risings and settings of Venus, but other celestial phenomena, such as lunar eclipses.

In those days, only a few Maya priests could do such elaborate calculations. Now, suppose we were to ask one of them how to do just one step in the process of predicting when Venus will next rise as a morning star—subtracting two numbers. And let's assume that, unlike today, we had not gone to school and did not know how to subtract. How would the priest explain to us what subtraction is?

He could either teach us the numbers represented by the bars and dots and the rules for "subtracting" them, or he could tell us what he was really doing: "Suppose we want to subtract 236 from 584. First, count out 584 beans and put them in a pot. Then take out 236 beans and put them to one side. Finally, count the beans left in the pot. That number is the result of subtracting 236 from 584."

You might say, "My Quetzalcoatl! What *tedium*—counting beans, putting them in, taking them out—what a job!"

To which the priest would reply, "That's why we have the rules for the bars and dots. The rules are tricky, but they are a much more efficient way of getting the answer than by counting beans. The important thing is, it makes no difference as far as the *answer* is concerned: we can predict the appearance of Venus by counting beans (which is slow, but easy to understand) or by using the tricky rules (which is much faster, but you must spend years in school to learn them)."

To understand *how* subtraction works—as long as you don't have to actually carry it out—is really not so difficult. That's my position: I'm going to explain to you what the physicists are *doing* when they are predicting how Nature will behave, but I'm not going to teach you any tricks so you can do it *efficiently*. You will discover that in order to make any reasonable predictions with this new scheme of quantum electrodynamics, you would have to make an awful lot of little arrows on a piece of paper. It takes seven years—four undergraduate and three graduate—to train our physics students to do that in a tricky, efficient way. That's where we are going to skip seven years of education in physics: By explaining quantum electrodynamics to you in terms of what we are *really doing*, I hope you will be able to understand it better than do some of the students!

Taking the example of the Maya one step further, we could ask the priest *why* five cycles of Venus nearly equal 2,920 days, or eight years. There would be all kinds of theories about *why*, such as, "20 is an important number in our counting system, and if you divide 2,920 by 20, you get 146, which is one more than a number that can be represented by the sum of two squares in two different ways," and so forth. But that theory would have nothing to do with Venus, really. In modern times, we have found that theories of this kind are not useful. So again, we are not going to deal with *why* Nature behaves in the peculiar way that She does; there are no good theories to explain that.

What I have done so far is to get you into the right mood to listen to me. Otherwise, we have no chance. So now we're off, ready to go!

We begin with light. When Newton started looking at light, the first thing he found was that white light is a mixture of colors. He separated white light with a prism into various colors, but when he put light of one color—red, for instance—through another prism, he found it could not be separated further. So Newton found that white light is a mixture of different colors, each of which is pure in the sense that it can't be separated further.

(In fact, a particular color of light can be split one more time in a different way, according to its so-called "polarization." This aspect of light is not vital to understanding the character of quantum electrodynamics, so for the sake of simplicity I will leave it out—at the expense of not giving you an absolutely complete description of the theory. This slight simplification will not remove, in any way, any real understanding of what I will be talking about. Still, I must be careful to mention all of the things I leave out.)

When I say "light" in these lectures, I don't mean simply the light we can see, from red to blue. It turns out that visible light is just a part of a long scale that's analogous to a musical scale in which there are notes higher than you can hear and other notes lower than you can hear. The scale of light can be described by numbers—called the frequency—and as the numbers get higher, the light goes from red to blue to violet to ultraviolet. We can't see ultraviolet light, but it can affect photographic plates. It's still light—only the number is different. (We shouldn't be so provincial: what we can detect directly with our own instrument, the eye, isn't the only thing in the world!) If we continue simply to change the number, we go out into X-rays, gamma rays, and so on. If we change the number in the other direction, we go from blue to red to infrared (heat) waves, then television waves, and radio waves. For me, all of that is "light." I'm going to use just red light for most of my examples, but the theory of quantum electrodynamics extends over the entire range I have described, and is the theory behind all these various phenomena.

Newton thought that light was made up of particles—he called them "corpuscles"—and he was right (but the reasoning that he used to come to that decision was erroneous). We know that light is made of particles because we can take a very sensitive instrument that makes clicks when light shines on it, and if the light gets dimmer, the clicks remain just as loud—there are just fewer of them. Thus light is something like raindrops—each little lump of light is called a photon—and if the light is all one color, all the "raindrops" are the same size.

The human eye is a very good instrument: it takes only about five or six photons to activate a nerve cell and send a message to the brain. If we were evolved a little further so we could see ten times more sensitively, we wouldn't have to have this discussion—we would all have seen very dim light of one color as a series of intermittent little flashes of equal intensity.

You might wonder how it is possible to detect a single photon. One instrument that can do this is called a photomultiplier, and I'll describe briefly how it works: When a photon hits the metal plate A at the bottom (see Figure 1), it causes an electron to break loose from one of the atoms in the plate. The free electron is strongly attracted to plate B (which has a positive charge on it) and hits it with enough force to break loose three or four electrons. Each of the electrons knocked out of plate B is attracted to plate C (which is also charged), and their collision with plate C knocks loose even more electrons. This process is repeated ten or twelve times, until billions of electrons, enough to make a sizable electric current, hit the last plate, L. This current can be amplified by a regular amplifier and sent through a speaker to make audible clicks. Each time a photon of a given color hits the photomultiplier, a click of uniform loudness is heard.

If you put a whole lot of photomultipliers around and let some very dim light shine in various directions, the light goes into one multiplier or another and makes a click of full intensity. It is all or nothing: if one photomultiplier

FIGURE 1. *A photomultiplier can detect a single photon. When a photon strikes plate A, an electron is knocked loose and attracted to positively charged plate B, knocking more electrons loose. This process continues until billions of electrons strike the last plate, L, and produce an electric current, which is amplified by a regular amplifier. If a speaker is connected to the amplifier, clicks of uniform loudness are heard each time a photon of a given color hits plate A.*

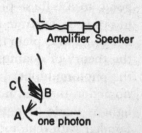

goes off at a given moment, none of the others goes off at the same moment (except in the rare instance that two photons happened to leave the light source at the same time). There is no splitting of light into "half particles" that go different places.

I want to emphasize that light comes in this form—particles. It is very important to know that light behaves like particles, especially for those of you who have gone to school, where you were probably told something about light behaving like waves. I'm telling you the way it *does* behave—like particles.

You might say that it's just the photomultiplier that detects light as particles, but no, every instrument that has been designed to be sensitive enough to detect weak light has always ended up discovering the same thing: light is made of particles.

I am going to assume that you are familiar with the properties of light in everyday circumstances—things like, light goes in straight lines; it bends when it goes into water; when it is reflected from a surface like a mirror, the angle at which the light hits the surface is equal to the angle at which it leaves the surface; light can be separated into colors; you can see beautiful colors on a mud puddle when there is a little bit of oil on it; a lens focuses light, and so on. I am going to use these phenomena that you are familiar with in order to illustrate the truly strange behavior of light; I am going to explain these familiar phenomena in terms of the theory of quantum electrodynamics. I told you about the photomultiplier in order to illustrate an essential phenomenon that you may not have been familiar with—that light is made of particles—but by now, I hope you are familiar with that, too!

Now, I think you are all familiar with the phenomenon that light is partly reflected from some surfaces, such as water. Many are the romantic paintings of moonlight reflecting from a lake (and many are the times you got yourself in trouble *because* of moonlight reflecting from a lake!). When you look down into water you can see what's below the surface (especially in the daytime), but you can also see a reflection from the surface. Glass is another example: if you have a lamp on in the room and you're looking out through a window during the daytime, you can see things outside through the glass as well as a dim reflection of the lamp in the room. So light is partially reflected from the surface of glass.

Before I go on, I want you to be aware of a simplification I am going to make that I will correct later on: When I talk about the partial reflection of light by glass, I am going to pretend that the light is reflected by only the *surface* of the glass. In reality, a piece of glass is a terrible monster of complexity—huge numbers of electrons are jiggling about.

When a photon comes down, it interacts with electrons *throughout* the glass, not just on the surface. The photon and electrons do some kind of dance, the net result of which is the same as if the photon hit only the surface. So let me make that simplification for a while. Later on, I'll show you what actually happens inside the glass so you can understand why the result is the same.

Now I'd like to describe an experiment, and tell you its surprising results. In this experiment some photons of the same color—let's say, red light—are emitted from a light source (see Fig. 2) down toward a block of glass. A photomultiplier is placed at A, above the glass, to catch any

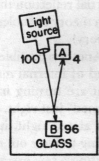

FIGURE 2. *An experiment to measure the partial reflection of light by a single surface of glass. For every 100 photons that leave the light source, 4 are reflected by the front surface and end up in the photomultiplier at A, while the other 96 are transmitted by the front surface and end up in the photomultiplier at B.*

photons that are reflected by the front surface. To measure how many photons get past the front surface, another photomultiplier is placed at B, inside the glass. Never mind the obvious difficulties of putting a photomultiplier inside a block of glass; what are the results of this experiment?

For every 100 photons that go straight down toward the glass at 90°, an average of 4 arrive at A and 96 arrive at B. So "partial reflection" in this case means that 4% of the photons are reflected by the front surface of the glass, while the other 96% are transmitted. *Already* we are in great difficulty: how can light be *partly* reflected? Each photon ends up at A or B—how does the photon "make up its mind" whether it should go to A or B? (Audience laughs.) That may sound like a joke, but we can't just laugh; we're going to have to explain that in terms of a theory! Partial reflection is already a deep mystery, and it was a very difficult problem for Newton.

There are several possible theories that you could make up to account for the partial reflection of light by glass. One of them is that 96% of the surface of the glass is "holes" that let the light through, while the other 4% of the surface is covered by small "spots" of reflective material (see Fig. 3). Newton realized that this is not a possible explanation.[1] In just a moment we will encounter a strange feature of partial reflection that will drive you crazy if you try to stick to a theory of "holes and spots"—or to any other reasonable theory!

Another possible theory is that the photons have some kind of internal mechanism—"wheels" and "gears" inside that are turning in some way—so that when a photon is "aimed" just right, it goes through the glass, and when it's not aimed right, it reflects. We can check this theory by trying to filter out the photons that are not aimed right by putting a few extra layers of glass between the source and the first layer of glass. After going through the filters, the photons reaching the glass should *all* be aimed right, and

[1] How did he know? Newton was a very great man: he wrote, "Because I can polish glass." You might wonder, how the heck could he tell that because you can polish glass, it can't be holes and spots? Newton polished his own lenses and mirrors, and he knew what he was doing with polishing: he was making scratches on the surface of a piece of glass with powders of increasing fineness. As the scratches become finer and finer, the surface of the glass changes its appearance from a dull grey (because the light is scattered by the large scratches), to a transparent clarity (because the extremely fine scratches let the light through). Thus he saw that it is impossible to accept the proposition that light can be affected by very small irregularities such as scratches or holes and spots; in fact, he found the contrary to be true. The finest scratches and therefore equally small spots do not affect the light. So the holes and spots theory is no good.

none of them should reflect. The trouble with that theory is, it doesn't agree with experiment: even after going through many layers of glass, 4% of the photons reaching a given surface reflect off it.

Try as we might to invent a reasonable theory that can

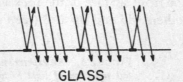

GLASS

FIGURE 3. *One theory to explain partial reflection by a single surface involves a surface made up mainly of "holes" that let light through, with a few "spots" that reflect the light.*

explain how a photon "makes up its mind" whether to go through glass or bounce back, it is impossible to predict which way a given photon will go. Philosophers have said that if the same circumstances don't always produce the same results, predictions are impossible and science will collapse. Here is a circumstance—identical photons are always coming down in the same direction to the same piece of glass—that produces different results. We cannot predict whether a given photon will arrive at A or B. All we can predict is that out of 100 photons that come down, an average of 4 will be reflected by the front surface. Does this mean that physics, a science of great exactitude, has been reduced to calculating only the *probability* of an event, and not predicting exactly what will happen? Yes. That's a retreat, but that's the way it is: Nature permits us to calculate only probabilities. Yet science has not collapsed.

While partial reflection by a single surface is a deep mystery and a difficult problem, partial reflection by two or more surfaces is absolutely mind-boggling. Let me

show you why. We'll do a second experiment, in which we
will measure the partial reflection of light by two surfaces.
We replace the block of glass with a very thin sheet of
glass—its two surfaces are exactly parallel to each other—
and we place the photomultiplier below the sheet of glass,
in line with the light source. This time, photons can reflect
from either the front surface or the back surface to end up
at A; all the others will end up at B (see Fig. 4). We might

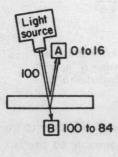

FIGURE 4. *An experiment to measure the partial reflection of light by two surfaces of glass.
Photons can get to the photomultiplier at A by
reflecting off either the front surface or the back
surface of the sheet of glass; alternatively, they
could go through both surfaces and end up hitting
the photomultiplier at B. Depending on the thickness of the glass, 0 to 16 photons out of every 100
get to the photomultiplier at A. These results pose
difficulties for any reasonable theory, including
the one in Figure 3. It appears that partial reflection can be "turned off" or "amplified" by the
presence of an additional surface.*

expect the front surface to reflect 4% of the light and the
back surface to reflect 4% of the remaining 96%, making a
total of about 8%. So we should find that out of every 100
photons that leave the light source, about 8 arrive at A.

What actually happens under these carefully controlled
experimental conditions is, the number of photons arriving
at A is rarely 8 out of 100. With some sheets of glass, we
consistently get a reading of 15 or 16 photons—twice our
expected result! With other sheets of glass, we consistently get
only 1 or 2 photons. Other sheets of glass have a partial reflection of 10%; some eliminate partial reflection altogether!
What can account for these crazy results? After checking
the various sheets of glass for quality and uniformity, we
discover that they differ only slightly in their thickness.

To test the idea that the amount of light reflected by two surfaces depends on the thickness of the glass, let's do a series of experiments: Starting out with the thinnest possible layer of glass, we'll count how many photons hit the photomultiplier at A each time 100 photons leave the light source. Then we'll replace the layer of glass with a slightly thicker one and make new counts. After repeating this process a few dozen times, what are the results?

With the thinnest possible layer of glass, we find that the number of photons arriving at A is nearly always zero—sometimes it's 1. When we replace the thinnest layer with a slightly thicker one, we find that the amount of light reflected is higher—closer to the expected 8%. After a few more replacements the count of photons arriving at A increases past the 8% mark. As we continue to substitute still "thicker" layers of glass—we're up to about 5 millionths of an inch now—the amount of light reflected by the two surfaces reaches a maximum of 16%, and then goes down, through 8%, back to zero—if the layer of glass is just the right thickness, there is no reflection at all. (Do *that* with spots!)

With gradually thicker and thicker layers of glass, partial reflection again increases to 16% and returns to zero—a cycle that repeats itself again and again (see Fig. 5). Newton discovered these oscillations and did one experiment that could be correctly interpreted only if the oscillations continued for at least 34,000 cycles! Today, with lasers (which produce a very pure, monochromatic light), we can see this cycle still going strong after more than 100,000,000 repetitions—which corresponds to glass that is more than 50 meters thick. (We don't see this phenomenon every day because the light source is normally not monochromatic.)

So it turns out that our prediction of 8% is right as an overall average (since the actual amount varies in a regular pattern from zero to 16%), but it's exactly right only twice each cycle—like a stopped clock (which is right twice a day).

Chapter 1

How can we explain this strange feature of partial reflection that depends on the thickness of the glass? How can the front surface reflect 4% of the light (as confirmed in our first experiment) when, by putting a second surface at just the right distance below, we can somehow "turn off" the

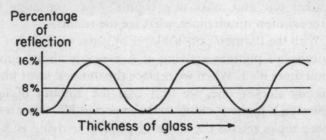

FIGURE 5. *The results of an experiment carefully measuring the relationship between the thickness of a sheet of glass and partial reflection demonstrate a phenomenon called "interference." As the thickness of the glass increases, partial reflection goes through a repeating cycle of zero to 16%, with no signs of dying out.*

reflection? And by placing that second surface at a slightly different depth, we can "amplify" the reflection up to 16%! Can it be that the back surface exerts some kind of influence or effect on the ability of the front surface to reflect light? What if we put in a *third* surface?

With a third surface, or any number of subsequent surfaces, the amount of partial reflection is again changed. We find ourselves chasing down through surface after surface with this theory, wondering if we have finally reached the last surface. Does a photon have to do that in order to "decide" whether to reflect off the front surface?

Newton made some ingenious arguments concerning this problem,[2] but he realized, in the end, that he had not yet developed a satisfactory theory.

  [2] It is very fortunate for us that Newton convinced himself that light is "corpuscles," because we can see what a fresh and intelligent mind looking

For many years after Newton, partial reflection by two surfaces was happily explained by a theory of waves,[3] but when experiments were made with very weak light hitting photomultipliers, the wave theory collapsed: as the light got dimmer and dimmer, the photomultipliers kept making

---

at this phenomenon of partial reflection by two or more surfaces has to go through to try to explain it. (Those who believed that light was waves never had to wrestle with it.) Newton argued as follows: Although light appears to be reflected from the first surface, it cannot be reflected from that surface. If it were, then how could light reflected from the first surface be captured again when the thickness is such that there was supposed to be no reflection at all? Then light must be reflected from the second surface. But to account for the fact that the thickness of the glass determines the amount of partial reflection, Newton proposed this idea: Light striking the first surface sets off a kind of wave or field that travels along with the light and predisposes it to reflect or not reflect off the second surface. He called this process "fits of easy reflection or easy transmission" that occur in cycles, depending on the thickness of the glass.

There are two difficulties with this idea: the first is the effect of additional surfaces—each new surface affects the reflection—which I described in the text. The other problem is that light certainly reflects off a lake, which doesn't have a second surface, so light *must* be reflecting off the front surface. In the case of single surfaces, Newton said that light had a predisposition to reflect. Can we have a theory in which the light knows what kind of surface it is hitting, and whether it is the only surface?

Newton didn't emphasize these difficulties with his theory of "fits of reflection and transmission," even though it is clear that he knew his theory was not satisfactory. In Newton's time, difficulties with a theory were dealt with briefly and glossed over—a different style from what we are used to in science today, where we point out the places where our own theory doesn't fit the observations of experiment. I'm not trying to say anything against Newton; I just want to say something in favor of how we communicate with each other in science today.

[3] This idea made use of the fact that waves can combine or cancel out, and the calculations based on this model matched the results of Newton's experiments, as well as those done for hundreds of years afterwards. But when instruments were developed that were sensitive enough to detect a single photon, the wave theory predicted that the "clicks" of the photomultiplier would get softer and softer, whereas they stayed at full strength—they just occurred less and less often. No reasonable model could explain this fact, so there was a period for a while in which you had to be clever: You had to know which experiment you were analyzing in order to tell if light was waves or particles. This state of confusion was called the "wave-particle duality" of light, and it was jokingly said by someone that light was waves on Mondays, Wednesdays, and Fridays; it was particles on Tuesdays, Thursdays, and Saturdays, and on Sundays, we think about it! It is the purpose of these lectures to tell you how this puzzle was finally "resolved."

full-sized clicks—there were just fewer of them. Light be-
haved as particles.

The situation today is, we haven't got a good model to
explain partial reflection by two surfaces; we just calculate
the probability that a particular photomultiplier will be hit
by a photon reflected from a sheet of glass. I have chosen
this calculation as our first example of the method provided
by the theory of quantum electrodynamics. I am going to
show you "how we count the beans"—what the physicists
do to get the right answer. I am not going to explain how
the photons actually "decide" whether to bounce back or
go through; that is not known. (Probably the question has
no meaning.) I will only show you how to calculate the
correct *probability* that light will be reflected from glass of
a given thickness, because that's the only thing physicists
know how to do! What we do to get the answer to *this*
problem is analogous to the things we have to do to get the
answer to *every other* problem explained by quantum
electrodynamics.

You will have to brace yourselves for this—not because
it is difficult to understand, but because it is absolutely
ridiculous: All we do is draw little arrows on a piece of
paper—that's all!

Now, what does an arrow have to do with the chance that
a particular event will happen? According to the rules of
"how we count the beans," the probability of an event is
equal to the square of the length of the arrow. For example,
in our first experiment (when we were measuring partial
reflection by the front surface only), the probability that a
photon would arrive at the photomultiplier at A was 4%.
That corresponds to an arrow whose length is 0.2, because
0.2 squared is 0.04 (see Fig. 6).

In our second experiment (when we were replacing thin
sheets of glass with slightly thicker ones), photons bouncing
off either the front surface or the back surface arrived at

A. How do we draw an arrow to represent this situation?
The length of the arrow must range from zero to 0.4 to
represent probabilities of zero to 16%, depending on the
thickness of the glass (see Fig. 7).

We start by considering the various *ways* that a photon

FIGURE 6. *The strange feature of partial reflection by
two surfaces has forced physicists away from making absolute predictions to merely calculating the probability of
an event. Quantum electrodynamics provides a method
for doing this—drawing little arrows on a piece of paper.
The probability of an event is represented by the area of
the square on an arrow. For example, an arrow representing a probability of 0.04 (4%) has a length of 0.2.*

could get from the source to the photomultiplier at A. Since
I am making this simplification that the light bounces off
either the front surface or the back surface, there are two
possible ways a photon could get to A. What we do in this
case is to draw *two* arrows—one for each way the event can
happen—and then combine them into a "final arrow"

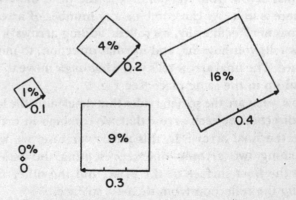

FIGURE 7. *Arrows representing probabilities from
0% to 16% have lengths of from 0 to 0.4.*

whose square represents the probability of the event. If there had been three different ways the event could have happened, we would have drawn three separate arrows before combining them.

Now, let me show you how we combine arrows. Let's say we want to combine arrow x with arrow y (see Fig. 8). All

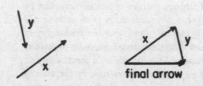

FIGURE 8. *Arrows that represent each possible way an event could happen are drawn and then combined ("added") in the following manner: Attach the head of one arrow to the tail of another—without changing the direction of either one—and draw a "final arrow" from the tail of the first arrow to the head of the last one.*

we have to do is put the head of x against the tail of y (without changing the direction of either one), and draw the final arrow from the tail of x to the head of y. That's all there is to it. We can combine any number of arrows in this manner (technically, it's called "adding arrows"). Each arrow tells you how far, and in what direction, to move in a dance. The final arrow tells you what *single* move to make to end up in the same place (see Fig. 9).

Now, what are the specific rules that determine the length and direction of each arrow that we combine in order to make the final arrow? In this particular case, we will be combining two arrows—one representing the reflection from the *front* surface of the glass, and the other representing the reflection from the *back* surface.

Let's take the length first. As we saw in the first experiment (where we put the photomultiplier inside the glass), the front surface reflects about 4% of the photons that come down. That means the "front reflection" arrow has a length of 0.2. The back surface of the glass also reflects 4%, so the "back reflection" arrow's length is also 0.2.

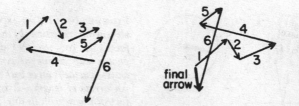

FIGURE 9. *Any number of arrows can be added in the manner described in Figure 8.*

To determine the direction of each arrow, let's imagine that we have a stopwatch that can time a photon as it moves. This imaginary stopwatch has a single hand that turns around very, very rapidly. When a photon leaves the source, we start the stopwatch. As long as the photon moves, the stopwatch hand turns (about 36,000 times per inch for red light); when the photon ends up at the photo-multiplier, we stop the watch. The hand ends up pointing in a certain direction. That is the direction we will draw the arrow.

We need one more rule in order to compute the answer correctly: When we are considering the path of a photon bouncing off the *front* surface of the glass, we reverse the direction of the arrow. In other words, whereas we draw the *back* reflection arrow pointing in the *same* direction as the stopwatch hand, we draw the *front* reflection arrow in the *opposite* direction.

Now, let's draw the arrows for the case of light reflecting from an extremely thin layer of glass. To draw the front reflection arrow, we imagine a photon leaving the light source (the stopwatch hand starts turning), bouncing off the front surface, and arriving at A (the stopwatch hand stops). We draw a little arrow of length 0.2 in the direction opposite that of the stopwatch hand (see Fig. 10).

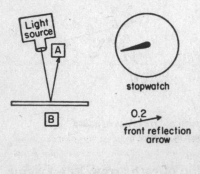

stopwatch

0.2
front reflection
arrow

FIGURE 10. *In an experiment measuring reflection by two surfaces, we can say that a single photon can arrive at A in two ways—via the front or back surface. An arrow of length 0.2 is drawn for each way, with its direction determined by the hand of a "stopwatch" that times the photon as it moves. The "front reflection" arrow is drawn in the direction opposite to that of the stopwatch hand when it stops turning.*

To draw the back reflection arrow, we imagine a photon leaving the light source (the stopwatch hand starts turning), going through the front surface and bouncing off the back surface, and arriving at A (the stopwatch hand stops). This time, the stopwatch hand is pointing in almost the same direction, because a photon bouncing off the back surface of the glass takes only slightly longer to get to A—it goes through the extremely thin layer of glass twice. We now draw a little arrow of length 0.2 in the same direction that the stopwatch hand is pointing (see Fig. 11).

Now let's combine the two arrows. Since they are both the same length but pointing in nearly opposite directions, the final arrow has a length of nearly zero, and its square is even closer to zero. Thus, the probability of light reflecting from an infinitesimally thin layer of glass is essentially zero (see Fig. 12).

FIGURE 11.   *A photon bouncing
off the back surface of a thin layer
of glass takes slightly longer to get
to A. Thus, the stopwatch hand ends
up in a slightly different direction
than it did when it timed the front
reflection photon. The "back reflection" arrow is drawn in the same
direction as the stopwatch hand.*

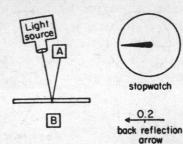

stopwatch

back reflection
arrow

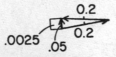

FIGURE 12.   *The final arrow, whose square represents the probability of
reflection by an extremely thin layer of glass, is drawn by adding the front
reflection arrow and the back reflection arrow. The result is nearly zero.*

When we replace the thinnest layer of glass with a slightly
thicker one, the photon bouncing off the back surface takes
a little bit longer to get to A than in the first example; the
stopwatch hand therefore turns a little bit more before it
stops, and the back reflection arrow ends up in a slightly
greater angle relative to the front reflection arrow. The
final arrow is a little bit longer, and its square is correspondingly larger (see Fig. 13).

As another example, let's look at the case where the glass
is just thick enough that the stopwatch hand makes an extra
half turn as it times a photon bouncing off the back surface.
This time, the back reflection arrow ends up pointing in
exactly the same direction as the front reflection arrow.
When we combine the two arrows, we get a final arrow
whose length is 0.4, and whose square is 0.16, representing
a probability of 16% (see Fig. 14).

If we increase the thickness of the glass just enough so

stopwatch

0.2 →

front reflection
arrow

stopwatch

0.2

back reflection
arrow

FIGURE 13. *The final arrow for a slightly thicker sheet of glass is a little longer, due to the greater relative angle between the front and back reflection arrows. This is because a photon boucing off the back surface takes a little longer to reach A, compared to the previous example.*

5%   0.2

0.2

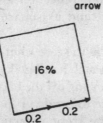

stopwatch

0.2 →

front reflection
arrow

stopwatch

0.2 →

back reflection
arrow

FIGURE 14. *When the layer of glass is just thick enough to allow the stopwatch hand timing the back reflecting photon to make an extra half turn, the front and back reflection arrows end up pointing in the same direction, resulting in a final arrow of length 0.4, which represents a probability of 16%.*

16%

0.2   0.2

that the stopwatch hand timing the back surface path makes an extra *full* turn, our two arrows end up pointing in opposite directions again, and the final arrow will be zero (see Fig. 15). This situation occurs over and over, whenever the thickness of the glass is just enough to let the stopwatch hand timing the back surface reflection make another full turn.

FIGURE 15. *When the sheet of glass is just the right thickness to allow the stopwatch hand timing the back reflecting photon to make one or more extra full turns, the final arrow is again zero, and there is no reflection at all.*

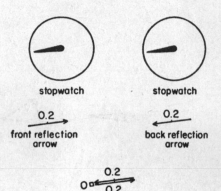

If the thickness of the glass is just enough to let the stopwatch hand timing the back surface reflection make an extra ¼ or ¾ of a turn, the two arrows will end up at right angles. The final arrow in this case is the hypotenuse of a right triangle, and according to Pythagoras, the square on the hypotenuse is equal to the sum of the squares on the other two sides. Here is the value that's right "twice a day"— 4% + 4% makes 8% (see Fig. 16).

Notice that as we gradually increase the thickness of the glass, the front reflection arrow always points in the same direction, whereas the back reflection arrow gradually changes its direction. The change in the relative direction of the two arrows makes the final arrow go through a re-

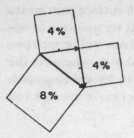

FIGURE 16. *When the front and back reflection arrows are at right angles to each other, the final arrow is the hypoteneuse of a right triangle. Thus its square is the sum of the other two squares—8%.*

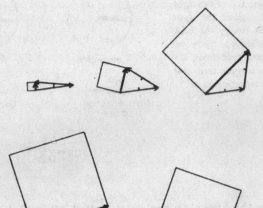

FIGURE 17. *As thin sheets of glass are replaced by slightly thicker ones, the stopwatch hand timing a photon reflecting off the back surface turns slightly more, and the relative angle between the front and back reflection arrows changes. This causes the final arrow to change in length, and its square to change in size from 0 to 16% back to 0, over and over.*

peating cycle of length zero to 0.4; thus the *square* on the final arrow goes through the repeating cycle of zero to 16% that we observed in our experiments (see Fig. 17).

I have just shown you how this strange feature of partial reflection can be accurately calculated by drawing some damned little arrows on a piece of paper. The technical word for these arrows is "probability amplitudes," and I feel more dignified when I say we are "computing the probability amplitude for an event." I prefer, though, to be more honest, and say that we are trying to find the arrow whose square represents the probability of something happening.

Before I finish this first lecture, I would like to tell you about the colors you see on soap bubbles. Or better, if your car leaks oil into a mud puddle, when you look at the brownish oil in that dirty mud puddle, you see beautiful colors on the surface. The thin film of oil floating on the mud puddle is something like a very thin sheet of glass— it reflects light of one color from zero to a maximum, depending on its thickness. If we shine pure red light on the film of oil, we see splotches of red light separated by narrow bands of black (where there's no reflection) because the oil film's thickness is not exactly uniform. If we shine pure blue light on the oil film, we see splotches of blue light separated by narrow bands of black. If we shine both red *and* blue light onto the oil, we see areas that have just the right thickness to strongly reflect only red light, other areas of the right thickness to reflect only blue light; still other areas have a thickness that strongly reflects both red and blue light (which our eyes see as violet), while other areas have the exact thickness to cancel out all reflection, and appear black.

To understand this better, we need to know that the cycle of zero to 16% partial reflection by two surfaces repeats more quickly for blue light than for red light. Thus at

certain thicknesses, one or the other or both colors are
strongly reflected, while at other thicknesses, reflection of
both colors is cancelled out (see Fig. 18). The cycles of
reflection repeat at different rates because the stopwatch
hand turns around faster when it times a blue photon than

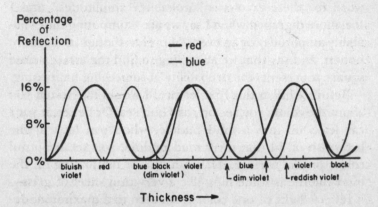

FIGURE 18. *As the thickness of a layer increases, the two surfaces produce
a partial reflection of monochromatic light whose probability fluctuates in a
cycle from 0% to 16%. Since the speed of the imaginary stopwatch hand is
different for different colors of light, the cycle repeats itself at different rates.
Thus when two colors such as pure red and pure blue are aimed at the layer,
a given thickness will reflect only red, only blue, both red and blue in different
proportions (which produce various hues of violet), or neither color (black).
If the layer is of varying thicknesses, such as a drop of oil spreading out on
a mud puddle, all of the combinations will occur. In sunlight, which consists
of all colors, all sorts of combinations occur, which produce lots of colors.*

it does when timing a red photon. In fact, that's the *only*
difference between a red photon and a blue photon (or a
photon of any other color, including radio waves, X-rays,
and so on)—the speed of the stopwatch hand.
    When we shine red and blue light on a film of oil, patterns
of red, blue, and violet appear, separated by borders of
black. When sunlight, which contains red, yellow, green,

and blue light, shines on a mud puddle with oil on it, the areas that strongly reflect each of those colors overlap and produce all kinds of combinations which our eyes see as different colors. As the oil film spreads out and moves over the surface of the water, changing its thickness in various locations, the patterns of color constantly change. (If, on the other hand, you were to look at the same mud puddle at night with one of those sodium streetlights shining on it, you would see only yellowish bands separated by black— because those particular streetlights emit light of only one color.)

This phenomenon of colors produced by the partial reflection of white light by two surfaces is called iridescence, and can be found in many places. Perhaps you have wondered how the brilliant colors of hummingbirds and peacocks are produced. Now you know. How those brilliant colors evolved is also an interesting question. When we admire a peacock, we should give credit to the generations of lackluster females for being selective about their mates. (Man got into the act later and streamlined the selection process in peacocks.)

In the next lecture I will show you how this absurd process of combining little arrows computes the right answer for those other phenomena you are familiar with: light travels in straight lines; it reflects off a mirror at the same angle that it came in ("the angle of incidence is equal to the angle of reflection"); a lens focuses light, and so on. This new framework will describe everything you know about light.

2

Photons:
Particles of Light

This is the second in a series of lectures about quantum electrodynamics, and since it's clear that none of you were here last time (because I told everyone that they weren't going to understand anything), I'll briefly summarize the first lecture.

We were talking about light. The first important feature about light is that it appears to be particles: when very weak monochromatic light (light of one color) hits a detector, the detector makes equally loud clicks less and less often as the light gets dimmer.

The other important feature about light discussed in the first lecture is partial reflection of monochromatic light. An average of 4% of the photons hitting a *single* surface of glass is reflected. This is already a deep mystery, since it is impossible to predict which photons will bounce back and which will go through. With a *second* surface, the results are strange: instead of the expected reflection of 8% by the two surfaces, the partial reflection can be amplified as high as 16% or turned off, depending on the thickness of the glass.

This strange phenomenon of partial reflection by two surfaces can be explained for intense light by a theory of waves, but the wave theory cannot explain how the detector makes equally loud clicks as the light gets dimmer. Quantum electrodynamics "resolves" this wave-particle duality by saying that light is made of particles (as Newton originally thought), but the price of this great advancement of science is a retreat by physics to the position of being able to calculate only the *probability* that a photon will hit a detector, without offering a good model of how it actually happens.

In the first lecture I described how physicists calculate the probability that a particular event will happen. They draw some arrows on a piece of paper according to some rules, which go as follows:

—GRAND PRINCIPLE: The probability of an event is equal to the square of the length of an arrow called the "probability amplitude." An arrow of length 0.4, for example, represents a probability of 0.16, or 16%.

—GENERAL RULE for drawing arrows if an event can happen in alternative ways: Draw an arrow for each way, and then combine the arrows ("add" them) by hooking the head of one to the tail of the next. A "final arrow" is then drawn from the tail of the first arrow to the head of the last one. The final arrow is the one whose square gives the probability of the entire event.

There were also some specific rules for drawing arrows in the case of partial reflection by glass (they can be found on pages 26 and 27).

All of the preceding is a review of the first lecture.

What I would like to do now is show you how this model of the world, which is so utterly different from anything you've ever seen before (that perhaps you hope never to see it again), can explain all the simple properties of light that you know: when light reflects off a mirror, the angle of incidence is equal to the angle of reflection; light bends when it goes from air into water; light goes in straight lines;

light can be focused by a lens, and so on. The theory also describes many other properties of light that you are probably not familiar with. In fact, the greatest difficulty I had in preparing these lectures was to resist the temptation to derive all of the things about light that took you so long to learn about in school—such as the behavior of light as it goes past an edge into a shadow (called diffraction)—but since most of you have not carefully observed such phenomena, I won't bother with them. However, I can guarantee you (otherwise, the examples I'm going to show you would be misleading) that *every* phenomenon about light that has been observed in detail can be explained by the theory of quantum electrodynamics, even though I'm going to describe only the simplest and most common phenomena.

We start with a mirror, and the problem of determining how light is reflected from it (see Fig. 19). At S we have a source that emits light of one color at very low intensity (let's use red light again). The source emits one photon at a time. At P, we place a photomultiplier to detect photons. Let's put it at the same height as the source—drawing arrows will be easier if everything is symmetrical. We want to calculate the chance that the detector will make a click after a photon has been emitted by the source. Since it is possible that a photon could go straight across to the detector, let's place a screen at Q to prevent that.

Now, we would expect that all the light that reaches the detector reflects off the middle of the mirror, because that's the place where the angle of incidence equals the angle of reflection. And it seems fairly obvious that the parts of the mirror out near the two ends have as much to do with the reflection as with the price of cheese, right?

Although you might *think* that the parts of the mirror near the two ends have nothing to do with the reflection of the light that goes from the source to the detector, let

(a)

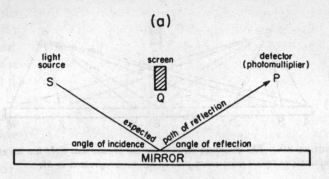

(b)

FIGURE 19. *The classical view of the world says that a mirror will reflect light where the angle of incidence is equal to the angle of reflection, even if the source and the detector are at different levels, as in (b).*

us look at what quantum theory has to say. Rule: The probability that a particular event occurs is the square of a final arrow that is found by drawing an arrow for each way the event could happen, and then combining ("adding") the arrows. In the experiment measuring the partial reflection of light by two surfaces, there were two ways a photon could get from the source to the detector. In this experiment, there are millions of ways a photon could go: it could go down to the left-hand part of the mirror at A or B (for example) and bounce up to the detector (see Fig. 20); it could bounce off the part where you think it should, at G; or, it could go down to the right-hand part of the mirror at K or M and bounce up to the detector. You might think

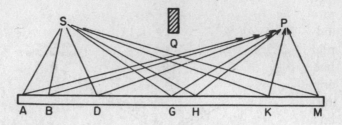

FIGURE 20. *The quantum view of the world says that light has an equal amplitude to reflect from every part of the mirror, from A to M.*

I'm crazy, because for most of the ways I told you a photon could reflect off the mirror, the angles aren't equal. But I'm *not* crazy, because that's the way light really goes! How can that be?

To make this problem easier to understand, let's suppose that the mirror consists of only a long strip from left to right—it's just as well that we forget, for a moment, that the mirror also sticks out from the paper (see Fig. 21). While

FIGURE 21. *To calculate more easily where the light goes, we shall temporarily consider only a strip of mirror divided into little squares, with one path for each square. This simplification in no way detracts from an accurate analysis of the situation.*

there are, in reality, millions of places where the light could reflect from this strip of mirror, let's make an approximation by temporarily dividing the mirror into a definite number of little squares, and consider only one path for each square—our calculation gets more accurate (but harder to do) as we make the squares smaller and consider more paths.

Now, let's draw a little arrow for each way the light could go in this situation. Each little arrow has a certain length and a certain direction. Let's consider the length first. You might think that the arrow we draw to represent the path that goes to the middle of the mirror, at G, is by far the longest (since there seems to be a very high probability that any photon that gets to the detector must go that way), and the arrows for the paths at the ends of the mirror must be very short. No, no; we should not make such an arbitrary rule. The right rule—what actually happens—is much simpler: a photon that reaches the detector has a nearly equal chance of going on *any* path, so all the little arrows have nearly the same length. (There are, in reality, some very slight variations in length due to the various angles and distances involved, but they are so minor that I am going to ignore them.) So let us say that each little arrow we draw will have an arbitrary standard length—I will make the length very short because there are many of these arrows representing the many ways the light could go (see Fig. 22).

FIGURE 22. *Each way the light can go will be represented in our calculation by an arrow of an arbitrary standard length, as shown.*

Although it is safe to assume that the length of all the arrows will be nearly the same, their directions will clearly differ because their timing is different—as you remember from the first lecture, the direction of a particular arrow is determined by the final position of an imaginary stopwatch that times a photon as it moves along that particular path. When a photon goes way off to the left end of the mirror, at A, and then up to the detector, it clearly takes more time than a photon that gets to the detector by reflecting in the middle of the mirror, at G (see Fig. 23). Or, imagine for a moment that you were in a hurry and had to run from the source over to the mirror and then to the detector. You'd know that it certainly isn't a good idea to go way over to A and then all the way up to the dectector; it would be much faster to touch the mirror somewhere in the middle.

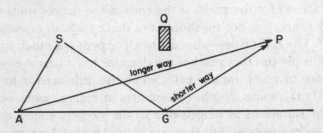

FIGURE 23. *While the length of each arrow is essentially the same, the direction will be different because the time it takes for a photon to go on each path is different. Clearly, it takes longer to go from S to A to P than from S to G to P.*

To help us calculate the direction of each arrow, I'm going to draw a graph right underneath my sketch of the mirror (see Fig. 24). Directly below each place on the mirror where the light could reflect, I'm going to show, vertically, how much time it would take if the light went that way. The more time it takes, the higher the point will be on the graph. Starting at the left, the time it takes a photon to go on the path that reflects at A is pretty long, so we plot a point pretty high up on the graph. As we move toward the center of the mirror, the time it takes for a photon to go the particular way we're looking at goes down, so we plot each successive point lower than the previous one. After we pass the center of the mirror, the time it takes a photon to go on each successive path gets longer and longer, so we plot our points correspondingly higher and higher. To aid the eye, let's connect the points: they form a symmetrical

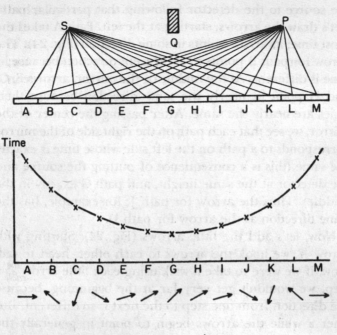

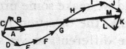

FIGURE 24. *Each path the light could go (in this simplified situation) is shown at the top, with a point on the graph below it showing the time it takes a photon to go from the source to that point on the mirror, and then to the photomultiplier. Below the graph is the direction of each arrow, and at the bottom is the result of adding all the arrows. It is evident that the major contribution to the final arrow's length is made by arrows E through I, whose directions are nearly the same because the timing of their paths is nearly the same. This also happens to be where the total time is least. It is therefore approximately right to say that light goes where the time is least.*

curve that starts high, goes down, and then goes back up again.

Now, what does that mean for the direction of the little arrows? The direction of a particular arrow corresponds to the amount of time it would take a photon to get from the source to the detector following that particular path. Let's draw the arrows, starting at the left. Path A takes the most time; its arrow points in some direction (Fig. 24). The arrow for path B points in a different direction because its time is different. At the middle of the mirror, arrows F, G, and H point in nearly the same direction because their times are nearly the same. After passing the center of the mirror, we see that each path on the right side of the mirror corresponds to a path on the left side whose time is exactly the same (this is a consequence of putting the source and the detector at the same height, and path G exactly in the middle). Thus the arrow for path J, for example, has the same direction as the arrow for path D.

Now, let's add the little arrows (Fig. 24). Starting with arrow A, we hook the arrows to each other, head to tail. Now, if we were to take a walk using each little arrow as a step, we wouldn't get very far at the beginning, because the direction from one step to the next is so different. But after a while the arrows begin to point in generally the same direction, and we make some progress. Finally, near the end of our walk, the direction from one step to the next is again quite different, so we stagger about some more.

All we have to do now is draw the final arrow. We simply connect the tail of the first little arrow to the head of the last one, and see how much direct progress we made on our walk (Fig. 24). And behold—we get a sizable final arrow! The theory of quantum electrodynamics predicts that light does, indeed, reflect off the mirror!

Now, let's investigate. What determines how long the final arrow is? We notice a number of things. First, the ends of the mirror are not important: there, the little arrows wander around and don't get anywhere. If I chopped off the ends of the mirror—parts that you instinctively knew I was wasting my time fiddling around with—it would hardly affect the length of the final arrow.

So where is the part of the mirror that gives the final arrow a substantial length? It's the part where the arrows are all pointing in nearly the same direction—because their *time* is almost the *same*. If you look at the graph showing the time for each path (Fig. 24), you see that the time is nearly the same from one path to the next at the bottom of the curve, where the *time* is *least*.

To summarize, where the time is least is also where the time for the nearby paths is nearly the same; that's where the little arrows point in nearly the same direction and add up to a substantial length; that's where the probability of a photon reflecting off a mirror is determined. And that's why, in approximation, we can get away with the crude picture of the world that says that light only goes where the *time* is *least* (and it's easy to prove that where the time is least, the angle of incidence is equal to the angle of reflection, but I don't have the time to show you).

So the theory of quantum electrodynamics gave the right answer—the middle of the mirror is the important part for reflection—but this correct result came out at the expense of believing that light reflects all over the mirror, and having to add a bunch of little arrows together whose sole purpose was to cancel out. All that might seem to you to be a waste of time—some silly game for mathematicians only. After all, it doesn't seem like "real physics" to have something there that only cancels out!

Let's test the idea that there really *is* reflection going on all over the mirror by doing another experiment. First, let's chop off most of the mirror, and leave about a quarter of it, over on the left. We still have a pretty big piece of mirror, but it's in the wrong place. In the previous experiment the arrows on the left side of the mirror were pointing in directions very different from one another because of the large difference in time between neighboring paths (Fig. 24). In this experiment I am going to make a more detailed calculation by taking intervals on that left-hand part of the mirror that are much closer together—fine enough that there is not much difference in time between neighboring paths (see Fig. 25). With this more detailed picture, we see that some of the arrows point more or less to the right; the others point more or less to the left. If we add *all* the arrows together, we have a bunch of arrows going around in what is essentially a circle, getting nowhere.

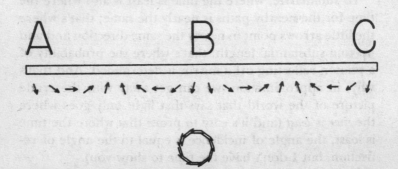

FIGURE 25. *To test the idea that there is really reflection happening at the ends of the mirror (but it is just cancelling out), we do an experiment with a large piece of mirror that is located in the wrong place for reflection from S to P. This piece of mirror is divided into much smaller sections, so that the timing from one path to the next is not very different. When all the arrows are added, they get nowhere: they go in a circle and add up to nearly nothing.*

But let's suppose we carefully scrape the mirror away in those areas whose arrows have a bias in one direction—let's say, to the left—so that only those places whose arrows point generally the other way remain (see Fig. 26). When we add up only the arrows that point more or less to the right, we get a series of dips and a substantial final arrow—according to the theory, we should now have a strong reflection! And indeed, we do—the theory *is* correct! Such a mirror is called a diffraction grating, and it works like a charm.

Isn't it wonderful—you can take a piece of mirror where

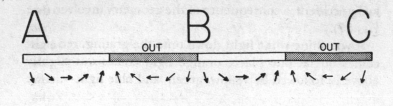

FIGURE 26. *If only the arrows with a bias in a particular direction—such as to the right—are added, while the others are disregarded (by etching away the mirror in those places), then a substantial amount of light reflects from this piece of mirror located in the wrong place. Such an etched mirror is called a diffraction grating.*

you didn't expect any reflection, scrape away part of it, and it reflects![1]

The particular grating that I just showed you was tailor-made for red light. It wouldn't work for blue light; we would have to make a new grating with the cut-away strips spaced closer together because, as I told you in the first lecture, the stopwatch hand turns around faster when it times a blue photon compared to a red photon. So the cuts that were especially designed for the "red" rate of turning don't fall in the right places for blue light; the arrows get kinked up and the grating doesn't work very well. But as a matter of accident, it happens that if we move the photomultiplier down to a somewhat different angle, the grating made for red light now works for blue light. It's just a

---

[1] The areas of the mirror whose arrows point generally to the left also make a strong reflection (when the areas whose arrows point the other way are erased). It's when both left-biased and right-biased areas reflect together that they cancel out. This is analogous to the case of partial reflection by two surfaces: while either surface will reflect on its own, if the thickness is such that the two surfaces contribute arrows pointing in opposite directions, reflection is cancelled out.

lucky accident, a consequence of the geometry involved (see Fig. 27).

If you shine white light down onto the grating, red light comes out at one place, orange light comes out slightly above it, followed by yellow, green, and blue light—all the

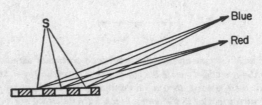

FIGURE 27. *A diffraction grating with grooves at the right distance for red light also works for other colors, if the detector is in a different place. Thus it is possible to see different colors reflecting from a grooved surface—such as a phonograph record—depending on the angle.*

colors of the rainbow. Where there is a series of grooves close together, you can often see colors—for example, when you hold a phonograph record (or better, a videodisc)— under bright light at the correct angles. Perhaps you have seen those wonderful silvery signs (here in sunny California they're often on the backs of cars): when the car moves, you see very bright colors changing from red to blue. Now you know where the colors come from: you're looking at a grating—a mirror that's been scratched in just the right places. The sun is the light source, and your eyes are the detector. I could go on to easily explain how lasers and holograms work, but I know that not everyone has seen these things, and I have too many other things to talk about.[2]

[2] I can't resist telling you about a grating that Nature has made: salt crystals are sodium and chlorine atoms packed in a regular pattern.

So a grating shows that we can't ignore the parts of a mirror that don't seem to be reflecting; if we do some clever things to the mirror, we can demonstrate the reality of the reflections from all parts of the mirror and produce some striking optical phenomena.

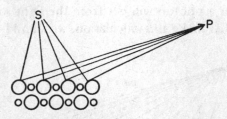

FIGURE 28. *Nature has made many types of diffraction gratings in the form of crystals. A salt crystal reflects X-rays (light for which the imaginary stopwatch hand moves extremely fast—perhaps 10,000 times faster than for visible light) at various angles, from which can be determined the exact arrangement and spacings of the individual atoms.*

More importantly, demonstrating the reality of reflection from *all* parts of the mirror shows that there is an amplitude—an arrow—for *every way* an event can happen. And in order to calculate correctly the probability of an event in different circumstances, we have to add the arrows for *every way* that the event could happen—not just the ways we think are the important ones!

Their alternating pattern, like our grooved surface, acts like a grating when light of the right color (X-rays, in this case) shines on it. By finding the specific locations where a detector picks up a lot of this special reflection (called diffraction), one can determine exactly how far apart the grooves are, and thus how far apart the atoms are (see Fig. 28). It is a beautiful way of determining the structure of all kinds of crystals as well as confirming that X-rays are the same thing as light. Such experiments were first done in 1914. It was very exciting to see, in detail, for the first time how the atoms are packed together in different substances.

Now, I would like to talk about something more familiar than gratings—about light going from air into water. This time, let's put the photomultiplier underwater—we suppose the experimenter can arrange that! The source of light is in the air at S, and the dectector is underwater, at D (see Fig. 29). Once again, we want to calculate the probability that a photon will get from the light source to the detector. To make this calculation, we should consider all

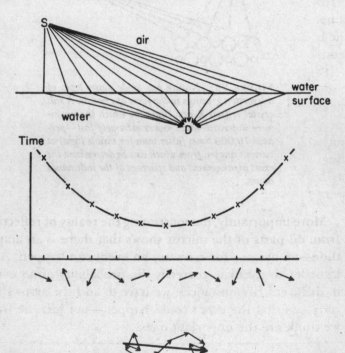

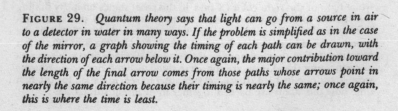

FIGURE 29. *Quantum theory says that light can go from a source in air to a detector in water in many ways. If the problem is simplified as in the case of the mirror, a graph showing the timing of each path can be drawn, with the direction of each arrow below it. Once again, the major contribution toward the length of the final arrow comes from those paths whose arrows point in nearly the same direction because their timing is nearly the same; once again, this is where the time is least.*

the ways the light could go. Each way the light could go contributes a little arrow and, as in the previous example, all the little arrows have nearly the same length. We can again make a graph of the time it takes a photon to go on each possible path. The graph will be a curve very similar to the one we made for light reflecting off a mirror: it starts up high, goes down, and then back up again; the most important contributions come from the places where the arrows point in nearly the same direction (where the time is nearly the same from one path to the next), which is at the bottom of the curve. That is also where the time is the least, so all we have to do is find out where the time is least.

It turns out that light seems to go slower in water than it does in air (I will explain why in the next lecture), which makes the distance through water more "costly," so to speak, than the distance through air. It's not hard to figure out which path takes the least time: suppose you're the lifeguard, sitting at S, and the beautiful girl is drowning, at D (Fig. 30). You can run on land faster than you can swim in water. The problem is, where do you enter the water in order to reach the drowning victim the fastest? Do you run down to the water at A, and then swim like

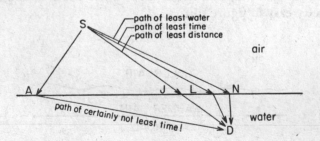

FIGURE 30. *Finding the path of least time for light is like finding the path of least time for a lifeguard running and then swimming to rescue a drowning victim: the path of least distance has too much water in it; the path of least water has too much land in it; the path of least time is a compromise between the two.*

hell? Of course not. But running directly toward the victim and entering the water at J is not the fastest route, either. While it would be foolish for a lifeguard to analyze and calculate under the circumstances, there is a computable position at which the time is minimum: it's a compromise between taking the direct path, through J, and taking the path with the least water, through N. And so it is with light—the path of least time enters the water at a point between J and N, such as L.

Another phenomenon of light that I would like to mention briefly is the mirage. When you're driving along a road that is very hot, you can sometimes see what looks like water on the road. What you're really seeing is the sky, and when you normally see sky on the road, it's because the road has puddles of water on it (partial reflection of light by a single surface). But how can you see sky on the road when there's no water there? What you need to know is that light goes slower through cooler air than through warmer air, and for a mirage to be seen, the observer must be in the cooler air that is above the hot air next to the road surface (see Fig. 31). How it is possible to look *down* and see the sky can be understood by finding the path of least time. I'll let you play with that one at home—it's fun to think about, and pretty easy to figure out.

FIGURE 31. *Finding the path of least time explains how a mirage works. Light goes faster through warm air than through cool air. Some of the sky appears to be on the road because some of the light from the sky reaches the eye by coming up from the road. The only other time sky appears to be on the road is when water is reflecting it, and thus a mirage appears to be water.*

In the examples I showed you of light reflecting off a mirror and of light going through air and then water, I was making an approximation: for the sake of simplicity, I drew the various ways the light could go as double straight lines—two straight lines that form an angle. But we don't have to *assume* that light goes in straight lines when it is in a uniform material like air or water; even *that* is explainable by the general principle of quantum theory: the probability of an event is found by adding arrows for *all* the ways the event could happen.

So for our next example, I'm going to show you how, by adding little arrows, it can appear that light goes in a straight line. Let's put a source and a photomultiplier at S and P, respectively (see Fig. 32), and look at *all* the ways

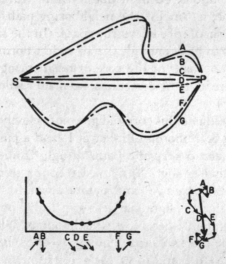

FIGURE 32. *Quantum theory can be used to show why light appears to travel in straight lines. When all possible paths are considered, each crooked path has a nearby path of considerably less distance and therefore much less time (and a substantially different direction for the arrow). Only the paths near the straight-line path at D have arrows pointing in nearly the same direction, because their timings are nearly the same. Only such arrows are important, because it is from them that we accumulate a large final arrow.*

the light could go—in all sorts of crooked paths—to get from the source to the detector. Then we draw a little arrow for each path, and we're learning our lesson well!

For each crooked path, such as path A, there's a nearby path that's a little bit straighter and distinctly shorter—that is, it takes much less time. But where the paths become nearly straight—at C, for example—a nearby, straighter path has nearly the same time. That's where the arrows add up rather than cancel out; that's where the light goes.

It is important to note that the single arrow that represents the straight-line path, through D (Fig. 32), is not enough to account for the probability that light gets from the source to the detector. The nearby, nearly straight paths—through C and E, for example—also make important contributions. So light doesn't *really* travel only in a straight line; it "smells" the neighboring paths around it, and uses a small core of nearby space. (In the same way, a mirror has to have enough size to reflect normally: if the mirror is too small for the core of neighboring paths, the light scatters in many directions, no matter where you put the mirror.)

Let's investigate this core of light more closely by putting a source at S, a photomultiplier at P, and a pair of blocks between them to keep the paths of light from wandering too far away (see Fig. 33). Now, let's put a second photo-multiplier at Q, below P, and assume again, for the sake of simplicity, that the light can get from S to Q only by paths of double straight lines. Now, what happens? When the gap between the blocks is wide enough to allow many neighboring paths to P and to Q, the arrows for the paths to P add up (because all the paths to P take nearly the same time), while the paths to Q cancel out (because those paths have a sizable difference in time). Thus the photomultiplier at Q doesn't click.

But as we push the blocks closer together, at a certain

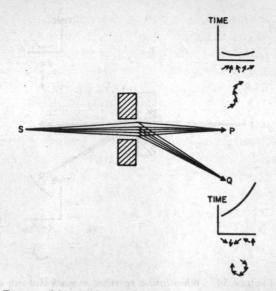

FIGURE 33. *Light travels in not just the straight-line path, but in the nearby paths as well. When two blocks are separated enough to allow for these nearby paths, the photons proceed normally to P, and hardly ever go to Q.*

point, the detector at Q starts clicking! When the gap is nearly closed and there are only a few neighboring paths, the arrows to Q *also* add up, because there is hardly any difference in time between them, either (see Fig. 34). Of course, both final arrows are small, so there's not much light either way through such a small hole, but the detector at Q clicks almost as much as the one at P! So when you try to squeeze light too much to make sure it's going in only a straight line, it refuses to cooperate and begins to spread out.[3]

[3] This is an example of the "uncertainty principle": there is a kind of "complementarity" between knowledge of where the light goes between the blocks and where it goes afterwards—precise knowledge of both is impossible. I would like to put the uncertainty principle in its historical place: When the revolutionary ideas of quantum physics were first coming

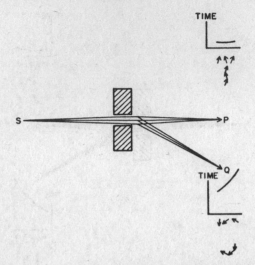

FIGURE 34. *When light is restricted so much that only a few paths are possible, the light that is able to get through the narrow slit goes to Q almost as much as to P, because there are not enough arrows representing the paths to Q to cancel each other out.*

So the idea that light goes in a straight line is a convenient approximation to describe what happens in the world that is familiar to us; it's similar to the crude approximation that says when light reflects off a mirror, the angle of incidence is equal to the angle of reflection.

Just as we were able to do a clever trick to make light reflect off a mirror at many angles, we can do a similar

---

out, people still tried to understand them in terms of old-fashioned ideas (such as, light goes in straight lines). But at a certain point the old-fashioned ideas would begin to fail, so a warning was developed that said, in effect, "Your old-fashioned ideas are no damn good when . . ." If you get rid of all the old-fashioned ideas and instead use the ideas that I'm explaining in these lectures—adding *arrows* for all the ways an event can happen—there is no need for an uncertainty principle!

trick to get light to go from one point to another in many ways.

First, to simplify the situation, I'm going to draw a vertical dashed line (see Fig. 35) between the light source and the detector (the line means nothing; it's just an artificial line)

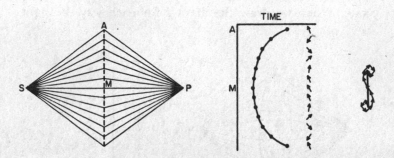

FIGURE 35.   *Analysis of all possible paths from S to P is simplified to include only double straight lines (in a single plane). The effect is the same as in the more complicated, real case: there is a time curve with a minimum, where most of the contribution to the final arrow is made.*

and say that the only paths we're going to look at are double straight lines. The graph that shows the time for each path looks the same as in the case of the mirror (but I'll draw it sideways, this time): the curve starts at A, at the top, and then it comes in, because the paths in the middle are shorter and take less time. Finally, the curve goes back out again.

Now, let's have some fun. Let's "fool the light," so that *all* the paths take exactly the same amount of time. How can we do this? How can we make the shortest path, through M, take exactly the same time as the longest path, through A?

Well, light goes slower in water than it does in air; it also goes slower in glass (which is much easier to handle!). So, if we put in just the right thickness of glass on the shortest path, through M, we can make the time for that path exactly the same as for the path through A. The paths next to M, which are just a little longer, won't need quite as much glass (see Fig. 36). The nearer we get to A, the less glass we have to put in to slow up the light. By carefully calculating and putting in just the right thickness of glass to compensate for the time along each path, we can make all the times the same. When we draw the arrows for each way the light

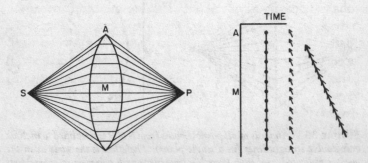

FIGURE 36. A "trick" can be played on Nature by slowing down the light that takes shorter paths: glass of just the right thickness is inserted so that all the paths will take exactly the same time. This causes all of the arrows to point in the same direction, and to produce a whopping final arrow—lots of light! Such a piece of glass made to greatly increase the probability of light getting from a source to a single point is called a focusing lens.

could go, we find we have succeeded in straightening them all out—and there are, in reality, *millions* of tiny arrows—so the net result is a sensationally large, unexpectedly enormous final arrow! Of course you know what I'm describing; it's a focusing lens. By arranging things so that all the times are equal, we can focus light—we can make the probability very high that light will arrive at a particular point, and very low that it will arrive anywhere else.

I have used these examples to show you how the theory of quantum electrodynamics, which looks at first like an absurd idea with no causality, no mechanism, and nothing real to it, produces effects that you are familiar with: light bouncing off a mirror, light bending when it goes from air into water, and light focused by a lens. It also produces other effects that you may or may not have seen, such as the diffraction grating and a number of other things. In fact, the theory continues to be successful at explaining *every* phenomenon of light.

I have shown you with examples how to calculate the probability of an event that can happen in *alternative ways*: we draw an arrow for each way the event can happen, and add the arrows. "Adding arrows" means the arrows are placed head to tail and a "final arrow" is drawn. The square of the resulting final arrow represents the probability of the event.

In order to give you a fuller flavor of quantum theory, I would now like to show you how physicists calculate the probability of compound events—events that can be broken down into a series of steps, or events that consist of a number of things happening independently.

An example of a compound event can be demonstrated by modifying our first experiment, in which we aimed some red photons at a single surface of glass to measure partial reflection. Instead of putting the photomultiplier at A (see Fig. 37), let's put in a screen with a hole in it to let the photons that reach point A go through. Then let's put in a sheet of glass at B, and place the photomultiplier at C. How do we figure out the probability that a photon will get from the source to C?

We can think of this event as a sequence of two steps. Step 1: a photon goes from the source to point A, reflecting off the single surface of glass. Step 2: the photon goes from point A to the photomultiplier at C, reflecting off the sheet of glass at B. Each step has a final arrow—an *"amplitude"* (I'm going to use the words interchangeably)—that can be calculated according to the rules we know so far. The amplitude for the first step has a length of 0.2 (whose square is 0.04, the probability of reflection by a single surface of glass), and is turned at some angle—let's say, 2 o'clock (Fig. 37).

To calculate the amplitude for the second step, we temporarily put the light source at A and aim the photons at the layer of glass above. We draw arrows for the front and back surface reflections and add them—let's say we end up with a final arrow with a length of 0.3, and turned toward 5 o'clock.

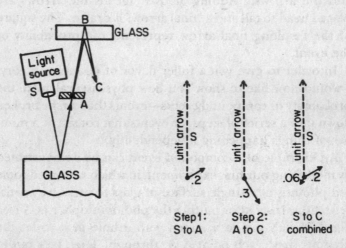

FIGURE 37. *A compound event can be analyzed as a succession of steps. In this example, the path of a photon going from S to C can be divided into two steps: 1) a photon gets from S to A, and 2) the photon gets from A to C. Each step can be analyzed separately to produce an arrow that can be regarded in a new way: as a unit arrow (an arrow of length 1 pointed at 12 o'clock) that has gone through a shrink and turn. In this example, the shrink and turn for Step 1 are 0.2 and 2 o'clock; the shrink and turn for Step 2 are 0.3 and 5 o'clock. To get the amplitude for the two steps in succession, we shrink and turn in succession: the unit arrow is shrunk and turned to produce an arrow of length 0.2 turned to 2 o'clock, which itself is shrunk and turned (as if it were the unit arrow) by 0.3 and 5 o'clock to produce an arrow of length 0.06 and turned to 7 o'clock. This process of successive shrinking and turning is called "multiplying" arrows.*

Now, how do we combine the two arrows to draw the amplitude for the entire event? We look at each arrow in a new way: as instructions for a *shrink* and *turn*.

In this example, the first amplitude has a length of 0.2 and is turned toward 2 o'clock. If we begin with a "unit arrow"—an arrow of length 1 pointed straight up—we can *shrink* this unit arrow from 1 down to 0.2, and *turn* it from 12 o'clock to 2 o'clock. The amplitude for the second step can be thought of as shrinking the unit arrow from 1 to 0.3 and turning it from 12 o'clock to 5 o'clock.

Now, to combine the amplitudes for both steps, we shrink and turn *in succession*. First, we shrink the unit arrow from 1 to 0.2 and turn it from 12 to 2 o'clock; then we shrink the arrow further, from 0.2 down to three-tenths of that, and turn it by the amount from 12 to 5—that is, we turn it from 2 o'clock to 7 o'clock. The resulting arrow has a length of 0.06 and is pointed toward 7 o'clock. It represents a probability of 0.06 squared, or 0.0036.

Observing the arrows carefully, we see that the result of shrinking and turning two arrows in succession is the same as adding their angles (2 o'clock + 5 o'clock) and multiplying their lengths (0.2 * 0.3). To understand why we add the angles is easy: the angle of an arrow is determined by the amount of turning by the imaginary stopwatch hand. So the total amount of turning for the two steps in succession is simply the sum of the turning for the first step plus the additional turning for the second step.

Why we call this process "multiplying arrows" takes a bit more explanation, but it's interesting. Let's look at multiplication, for a moment, from the point of view of the Greeks (this has nothing to do with the lecture). The Greeks wanted to use numbers that were not necessarily integers, so they represented numbers with lines. Any number can be expressed as a *transformation* of the unit line—by expanding it or shrinking it. For example, if Line A is the unit line (see Fig. 38), then line B represents 2 and line C represents 3.

Now, how do we multiply 3 times 2? We apply the transformations *in succession*: starting with line A as the unit line, we expand it 2 times and then 3 times (or 3 times and then 2 times—the order doesn't make any difference). The result is line D, whose length represents 6. What about multiplying 1/3 times 1/2? Taking line D to be the unit line, now, we shrink it to 1/2 (line C) and then to 1/3 of that. The result is line A, which represents 1/6.

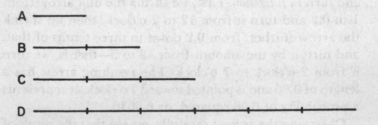

FIGURE 38. *We can express any number as a transformation of the unit line through expansion or shrinkage. If A is the unit line, then B represents 2 (expansion), and C represents 3 (expansion). Multiplying lines is achieved through successive transformations. For example, multiplying 3 by 2 means that the unit line is expanded 3 times and then 2 times, producing the answer, an expansion of 6 (line D). If D is the unit line, then line C represents 1/2 (shrinkage), line B represents 1/3 (shrinkage), and multiplying 1/2 by 1/3 means the unit line D is shrunk to 1/2, and then to 1/3 of that, producing the answer, a shrinkage to 1/6 (line A).*

Multiplying arrows works the same way (see Fig. 39). We apply transformations to the unit arrow in succession—it just happens that the transformation of an *arrow* involves *two* operations, a shrink and turn. To multiply arrow V times arrow W, we shrink and turn the unit arrow by the prescribed amounts for V, and then shrink it and turn it the amounts prescribed for W—again, the order doesn't make any difference. So multiplying arrows follows the

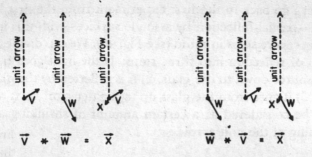

FIGURE 39. *Mathematicians found that multiplying arrows can also be expressed as successive transformations (for our purposes, successive shrinks and turns) of the unit arrow. As. in normal multiplication, the order is not important: the answer, arrow X, can be obtained by multiplying arrow V by arrow W or arrow W by arrow V.*

same rule of successive transformations that work for regular numbers.[4]

[4] Mathematicians have tried to find all the objects one could possibly find that obey the rules of algebra (A + B = B + A, A * B = B * A, and so on). The rules were originally made for positive integers, used for counting things like apples or people. Numbers were improved with the invention of zero, fractions, irrational numbers—numbers that cannot be expressed as a ratio of two integers—and negative numbers, and continued to obey the original rules of algebra. Some of the numbers that mathematicians invented posed difficulties for people at first—the idea of half a person was difficult to imagine—but today, there's no difficulty at all: nobody has any moral qualms or discomforting gory feelings when they hear that there is an average of 3.2 people per square mile in some regions. They don't try to imagine the 0.2 people; rather, they know what 3.2 means: if they multiply 3.2 by 10, they get 32. Thus, some things that satisfy the rules of algebra can be interesting to mathematicians even though they don't always represent a real situation. Arrows on a plane can be "added" by putting the head of one arrow on the tail of another, or "multiplied" by successive turns and shrinks. Since these arrows obey the same rules of algebra as regular numbers, mathematicians call them numbers. But to distinguish them from ordinary numbers, they're called "complex numbers." For those of you who have studied mathematics enough to have come to complex numbers, I could have said, "the probability of an event is the absolute square of a complex number. When an event can happen in alternative ways, you add the complex numbers; when it can happen only as a succession of steps, you multiply the complex numbers." Although it may sound more impressive that way, I have not said any more than I did before—I just used a different language.

64 Chapter 2

Let's go back to the first experiment from the first lecture—partial reflection by a single surface—with this idea of successive steps in mind (see Fig. 40). We can divide the path of reflection into three steps: 1) the light goes from the source down to the glass, 2) it is reflected by the glass, and 3) it goes from the glass up to the detector. Each step can be considered as a certain amount of shrinking and turning of the unit arrow.

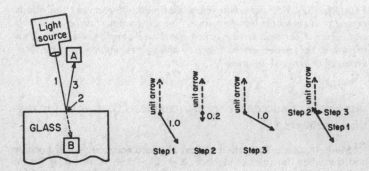

FIGURE 40. *Reflection by a single surface can be divided into three steps, each with a shrink and/or turn of the unit arrow. The net result, an arrow of length 0.2 pointed in some direction, is the same as before, but our method of analysis is more detailed now.*

You'll remember that in the first lecture, we did not consider *all* of the ways the light could reflect off the glass, which requires drawing and adding lots and lots of little tiny arrows. In order to avoid all that detail, I gave the impression that the light goes down to a particular point on the surface of the glass—that it doesn't spread out. When light goes from one point to another, it does, in reality, spread out (unless it's fooled by a lens), and there is some shrinkage of the unit arrow associated with that. For the moment, however, I would like to stick to the simplified view that light does *not* spread out, and so it is appropriate to disregard this shrinkage. It is also appropriate to assume that since the light doesn't spread out, every photon that leaves the source ends up at either A or B.

So: in the first step there is no shrinking, but there is turning—it corresponds to the amount of turning by the imaginary stopwatch hand as it times the photon going from the source to the front surface of the glass. In this example, the arrow for the first step ends up with a length of 1 at some angle—let's say, 5 o'clock.

The second step is the reflection of the photon by the glass. Here, there is a sizable shrink—from 1 to 0.2—and half a turn. (These numbers seem arbitrary now: they depend upon whether the light is reflected by glass or some other material. In the third lecture, I'll explain them, too!) Thus the second step is represented by an amplitude of length 0.2 and a direction of 6 o'clock (half a turn).

The last step is the photon going from the glass up to the detector. Here, as in the first step, there is no shrinking, but there is turning—let's say this distance is slightly shorter than in step 1, and the arrow points toward 4 o'clock.

We now "multiply" arrows 1, 2, and 3 in succession (add the angles, and multiply the lengths). The net effect of the three steps—1) turning, 2) a shrink and half a turn, and 3) turning—is the same as in the first lecture: the turning from steps 1 and 3—(5 o'clock plus 4 o'clock) is the same amount of turning that we got then when we let the stopwatch run for the whole distance (9 o'clock); the extra half turn from step 2 makes the arrow point in the direction opposite the stopwatch hand, as it did in the first lecture, and the shrinking to 0.2 in the second step leaves an arrow whose square represents the 4% partial reflection observed for a single surface.

In this experiment, there is a question we didn't look at in the first lecture: what about the photons that go to B— the ones that are transmitted by the surface of the glass?

Chapter 2

The amplitude for a photon to arrive at B must have a length near 0.98, since 0.98 * 0.98 = 0.9604, which is close enough to 96%. This amplitude can also be analyzed by breaking it down into steps (see Fig. 41).

The first step is the same as for the path to A—the photon goes from the light source down to the glass—the unit arrow is turned toward 5 o'clock.

The second step is the photon passing through the surface of the glass: there is no turning associated with transmission, just a little bit of shrinking—to 0.98.

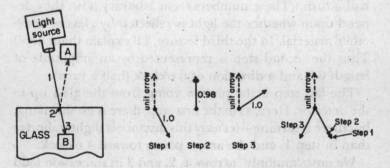

FIGURE 41. *Transmission by a single surface can also be divided into three steps, with a shrink and/or turn for each step. An arrow of length 0.98 has a square of about 0.96, representing a probabilty of transmission of 96% (which, combined with the 4% probability of reflection, accounts for 100% of the light).*

The third step—the photon going through the interior of the glass—involves additional turning and no shrinking.

The net result is an arrow of length 0.98 turned in some direction, whose square represents the probability that a photon will arrive at B—96%.

Now let's look at partial reflection by two surfaces again. Reflection from the front surface is the same as for a single surface, so the three steps for front surface reflection are the same as we saw a moment ago (Fig. 40).

Reflection from the back surface can be broken down into seven steps (see Fig. 42). It involves turning equal to the total amount of turning of the stopwatch hand timing a photon over the entire distance (steps 1, 3, 5, and 7), shrinking to 0.2 (step 4), and two shrinks to 0.98 (steps 2 and 6). The resulting arrow ends up in the same direction as before, but the length is about 0.192 (0.98 * 0.2 * 0.98), which I approximated as 0.2 in the first lecture.

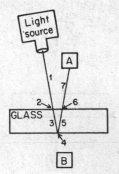

FIGURE 42. *Reflection from the back surface of a layer of glass can be divided into seven steps. Steps 1, 3, 5, and 7 involve turning only; steps 2 and 6 involve shrinks to 0.98, and step 4 involves a shrink to 0.2. The result is an arrow of length 0.192—which was approximated as 0.2 in the first lecture—turned at an angle that corresponds to the total amount of turning by the imaginary stopwatch hand.*

In summary, here are the rules for reflection and transmission of light by glass: 1) reflection from air back to air (off a front surface) involves a shrink to 0.2 and half a turn; 2) reflection from glass back to glass (off a back surface) also involves a shrink to 0.2, but no turning; and 3) transmission from air to glass or from glass to air involves a shrink to 0.98 and no turning in either case.

Perhaps it is too much of a good thing, but I cannot resist showing you a cute further example of how things work and are analyzed by these rules of successive steps. Let us move the detector to a location below the glass, and consider something we didn't talk about in the first lecture—the probability of *transmission* by two surfaces of glass (see Fig. 43).

Of course you know the answer: the probability of a

photon to arrive at B is simply 100% minus the probability to arrive at A, which we worked out beforehand. Thus, if we found the chance to arrive at A is 7%, the chance to arrive at B must be 93%. And as the chance for A varies from zero through 8% to 16% (due to the different thick-nesses of glass), the chance for B changes from 100% through 92% to 84%.

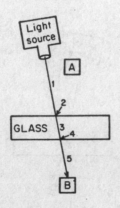

FIGURE 43. *Transmission by two surfaces can be broken down into five steps. Step 2 shrinks the unit arrow to 0.98, step 4 shrinks the 0.98 arrow to 0.98 of that (about 0.96); steps 1, 3, and 5 involve turning only. The resulting arrow of length 0.96 has a square of about 0.92, representing a probability of transmission by two surfaces of 92% (which corresponds to the expected 8% reflection, which is right only "twice a day"). When the thickness of the layer is right to produce a probability of 16% reflection, with a 92% probability of transmission, 108% of the light is accounted for! Something is wrong with this analysis!*

That is the right answer, but we are expecting to calculate *all* probabilities by squaring a final arrow. How do we calculate the amplitude arrow for transmission by a layer of glass, and how does it manage to vary in length so appropriately as to fit with the length for A in each case, so the probability for A and the probability for B always add up to exactly 100%? Let us look a little into the details.

For a photon to go from the source to the detector below the glass, at B, five steps are involved. Let's shrink and turn the unit arrow as we go along.

The first three steps are the same as in the previous example: the photon goes from the source to the glass (turning, no shrinking); the photon is transmitted by the front surface (no turning, shrinking to 0.98); the photon goes through the glass (turning, no shrinking).

The fourth step—the photon passes through the back surface of the glass—is the same as the second step, as far as shrinks and turns go: no turns, but a shrinkage to 0.98 of the 0.98, so the arrow now has a length of 0.96.

Finally, the photon goes through the air again, down to the detector—that means more turning, but no further shrinking. The result is an arrow of length 0.96, pointing in some direction determined by the successive turnings of the stopwatch hand.

An arrow whose length is 0.96 represents a probability of about 92% (0.96 squared), which means an average of 92 photons reach B out of every 100 that leave the source. That also means that 8% of the photons are reflected by the two surfaces and reach A. But we found out in the first lecture that an 8% reflection by two surfaces is only right sometimes ("twice a day")—that in reality, the reflection by two surfaces fluctuates in a cycle from zero to 16% as the thickness of the layer steadily increases. What happens when the glass is just the right thickness to make a partial reflection of 16%? For every 100 photons that leave the source, 16 arrive at A and 92 arrive at B, which means 108% of the light has been accounted for—horrifying! Something is wrong.

We neglected to consider *all* the ways the light could get to B! For instance, it could bounce off the back surface, go up through the glass as if it were going to A, but then reflect off the front surface, back down toward B (see Fig. 44). This path takes nine steps. Let's see what happens successively to the unit arrow as the light goes through each step (don't worry; it only shrinks and turns!).

First step—photon goes through the air—turning; no shrinking. Second step—photon passes through the glass—no turning, but shrinking to 0.98. Third step—photon goes through the glass—turning; no shrinking. Fourth step—reflection off the back surface—no turning, but shrinking to 0.2 of 0.98, or 0.196. Fifth step—photon goes back up through the glass—turning; no shrinking. Sixth step—photon bounces off front surface (it's really a "back" surface, because the photon stays *inside* the glass)—no turning, but shrinking to 0.2 of 0.196, or 0.0392. Seventh step—photon

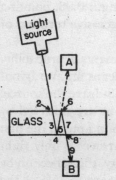

FIGURE 44. *Another way that light could be transmitted by two surfaces must be considered in order to make the calculation more accurate. This path involves two shrinks of 0.98 (steps 2 and 8) and two shrinks of 0.2 (steps 4 and 6), resulting in an arrow of length 0.0384 (rounded off to 0.04).*

goes back down through glass—more turning; no shrinking. Eighth step—photon passes through back surface—no turning, but shrinking to 0.98 of 0.0392, or 0.0384. Finally, the ninth step—photon goes through air to detector—turning; no shrinking.

The result of all this shrinking and turning is an amplitude of length 0.0384—call it 0.04, for all practical purposes—and turned at an angle that corresponds to the total amount of turning by the stopwatch as it times the photon going through this longer path. This arrow represents a *second* way that light can get from the source to B. Now we have two alternatives, so we must *add* the two arrows—the arrow for the more direct path, whose length is 0.96, and the arrow for the longer way, whose length is 0.04—to make the final arrow.

The two arrows are usually not in the same direction, because changing the thickness of the glass changes the relative direction of the 0.04 arrow to the 0.96 arrow. But look how nicely things work out: the extra turns made by the stopwatch timing a photon during steps 3 and 5 (on its way to A) are exactly equal to the extra turns it makes timing a photon during steps 5 and 7 (on its way to B). That means when the two reflection arrows are cancelling each other to make a final arrow representing zero reflection, the arrows for transmission are reinforcing each other to make an arrow of length 0.96 + 0.04, or 1—when the probability of reflection is zero, the probability of transmission is 100% (see Fig. 45). And when the arrows for reflection are rein-

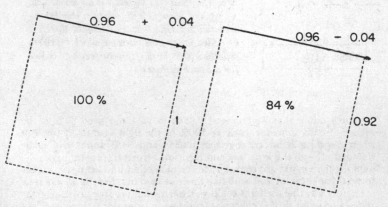

FIGURE 45. *Nature always makes sure 100% of the light is accounted for. When the thickness is right for the transmission arrows to accumulate, the arrows for reflection oppose each other; when the arrows for reflection accumulate, the arrows for transmission oppose each other.*

forcing each other to make an amplitude of 0.4, the arrows for transmission are going against each other, making an amplitude of length 0.96 − 0.04, or 0.92—when reflection is calculated to be 16%, transmission is calculated to be 84% (0.92 squared). You see how clever Nature is with Her rules to make sure that we always come out with 100% of the photons accounted for![5]

Finally, before I go, I would like to tell you that there is an extension to the rule that tells us when to multiply arrows: arrows are to be multiplied not only for an event that consists of a succession of steps, but also for an event that consists of a number of things happening concomitantly— independently and possibly simultaneously. For example, suppose we have two sources, X and Y, and two detectors, A and B (see Fig. 47), and we want to calculate the prob-

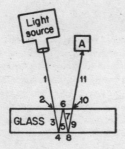

FIGURE 46. *Yet other ways the light could reflect should be considered for a more accurate calculation. In this figure, shrinks of 0.98 occur at steps 2 and 10; shrinks of 0.2 occur at steps 4, 6, and 8. The result is an arrow with a length of about 0.008, which is another alternative for reflection, and should therefore be added to the other arrows which represent reflection (0.2 for the front surface and 0.192 for the back surface).*

[5] You'll notice that we changed 0.0384 to 0.04 and used 84% as the square of 0.92, in order to make 100% of the light accounted for. But when *everything* is added together, 0.0384 and 84% don't have to be rounded off—all the little bits and pieces of arrows (representing all the ways the light could go) compensate for each other and keep the answer correct. For those of you who like this sort of thing, here is an example of another way that the light could go from the light source to the detector at A—a series of three reflections (and two transmissions), resulting in a final arrow of length 0.98 * 0.2 * 0.2 * 0.2 * 0.98, or about 0.008—a very tiny arrow (see Fig. 46). To make a complete calculation of partial reflection by two surfaces, you would have to add in that small arrow, plus an even smaller one that represents five reflections, and so on.

ability for the following event: after X and Y each lose a photon, A and B each gain a photon.

In this example, the photons travel through space to get to the detectors—they are neither reflected nor transmitted—so now is a good time for me to stop disregarding the fact that light spreads out as it goes along. I now present

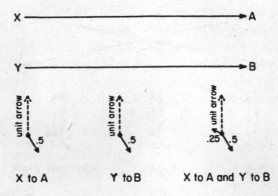

X to A   Y to B   X to A and Y to B

FIGURE 47. *If one of the ways a particular event can happen depends on a number of things happening independently, the amplitude for this way is calculated by multiplying the arrows of the independent things. In this case, the final event is: after sources X and Y each lose a photon, photomultipliers A and B make a click. One way this event could happen is that a photon could go from X to A and a photon could go from Y to B (two independent things). To calculate the probability for this "first way," the arrows for each independent thing—X to A and Y to B—are multiplied to produce the amplitude for this particular way. (Analysis continued in Fig. 48.)*

you with the *complete rule* for monochromatic light travelling from one point to another through space—there is nothing approximate here, and no simplification. This is all there is to know about monochromatic light going through space (disregarding polarization): the *angle* of the arrow depends on the imaginary stopwatch hand, which rotates a certain number of times per inch (depending on the color of the photon); the *length* of the arrow is inversely proportional to the distance the light goes—in other words, the arrow shrinks as the light goes along.[6]

Let's suppose the arrow for X to A is 0.5 in length and is pointing toward 5 o'clock, as is the arrow for Y to B (Fig. 47). Multiplying one arrow by the other, we get a final arrow of length 0.25, pointed at 10 o'clock.

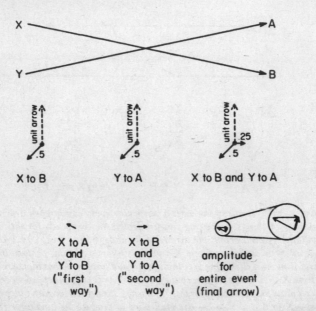

FIGURE 48. *The other way the event described in Figure 47 could happen— a photon goes from X to B and a photon goes from Y to A—also depends on two independent things happening, so the amplitude for this "second way" is also calculated by multiplying the arrows of the independent things. The "first way" and "second way" arrows are ultimately added together, resulting in the final arrow for the event. The probability of an event is always represented by a single final arrow—no matter how many arrows were drawn, multiplied, and added to achieve it.*

[6] This rule checks out with what they teach in school—the amount of light transmitted over a distance varies inversely as the square of the distance—because an arrow that shrinks to half its original size has a square one-fourth as big.

But wait! There is another way this event could happen: the photon from X could go to B, and the photon from Y could go to A. Each of these subevents has an amplitude, and these arrows must also be drawn and multiplied to produce an amplitude for this particular way the event could happen (see Fig. 48). Since the amount of shrinkage over distance is very small compared to the amount of turning, the arrows from X to B and Y to A have essentially the same length as the other arrows, 0.5, but their turning is quite different: the stopwatch hand rotates 36,000 times per inch for red light, so even a tiny difference in distance results in a substantial difference in timing.

The amplitudes for each way the event could happen are added to produce the final arrow. Since their lengths are essentially the same, it is possible for the arrows to cancel each other out if their directions are opposed to each other. The relative directions of the two arrows can be changed by changing the distance between the sources or the detectors: simply moving the detectors apart or together a little bit can make the probability of the event amplify or completely cancel out, just as in the case of partial reflection by two surfaces:[7]

In this example, arrows were multiplied and then added to produce a final arrow (the amplitude for the event), whose square is the probability of the event. It is to be emphasized that no matter how many arrows we draw, add, or multiply, our objective is to calculate a *single final arrow for the event*. Mistakes are often made by physics students at first because they do not keep this important point in mind. They work for so long analyzing events involving a single photon that they begin to think that the arrow is

[7] This phenomenon, called the Hanbury-Brown-Twiss effect, has been used to distinguish between a single source and a double source of radio waves in deep space, even when the two sources are extremely close together.

somehow associated with the photon. But these arrows are probability amplitudes, that give, when squared, the *probability* of a complete event.[8]

In the next lecture I will begin the process of simplifying and explaining the properties of matter—to explain where the shrinking to 0.2 comes from, why light appears to go slower through glass or water than through air, and so on— because I have been cheating so far: the photons don't really bounce off the surface of the glass; they interact with the electrons *inside* the glass. I'll show you how photons do nothing but go from one electron to another, and how reflection and transmission are really the result of an electron picking up a photon, "scratching its head," so to speak, and emitting a *new* photon. This simplification of everything we have talked about so far is very pretty.

[8] Keeping this principle in mind should help the student avoid being confused by things such as the "reduction of a wave packet" and similar magic.

3

Electrons and Their Interactions

This is the third of four lectures on a rather difficult subject—the theory of quantum electrodynamics—and since there are obviously more people here tonight than there were before, some of you haven't heard the other two lectures and will find this lecture almost incomprehensible. Those of you who *have* heard the other two lectures will also find this lecture incomprehensible, but you know that that's all right: as I explained in the first lecture, the way we have to describe Nature is generally incomprehensible to us.

In these lectures I want to tell you about the part of physics that we know best, the interaction of light and electrons. Most of the phenomena you are familiar with involve the interaction of light and electrons—all of chemistry and biology, for example. The only phenomena that are not covered by this theory are phenomena of gravitation and nuclear phenomena; everything else is contained in this theory.

We found out in the first lecture that we have no satisfactory mechanism to describe even the simplest of phenomena, such as partial reflection of light by glass. We also have no way to predict whether a given photon will be reflected or transmitted by the glass. All we can do is calculate the *probability* that a particular event will happen—

whether the light will be reflected, in this case. (This is about 4%, when the light shines straight down on a single surface of glass; the probability of reflection increases as the light hits the glass at more of a slant.)

When we deal with probabilities under *ordinary* circumstances, there are the following "rules of composition": 1) if something can happen in *alternative ways*, we *add* the probabilities for each of the different ways; 2) if the event occurs as a *succession of steps*—or depends on a number of things happening "concomitantly" (independently)—then we *multiply* the probabilities of each of the steps (or things).

In the wild and wonderful world of quantum physics, probabilities are calculated as the *square of the length of an arrow*: where we would have expected to add the probabilities under ordinary circumstances, we find ourselves "adding" *arrows*; where we normally would have multiplied the probabilities, we "multiply" *arrows*. The peculiar answers that we get from calculating probabilities in this manner match perfectly the results of experiment. I'm rather delighted that we must resort to such peculiar rules and strange reasoning in order to understand Nature, and I enjoy telling people about it. There are no "wheels and gears" beneath this analysis of Nature; if you want to understand Her, this is what you have to take.

Before I go into the main part of this lecture, I'd like to show you another example of how light behaves. What I would like to talk about is very weak light of one color—one photon at a time—going from a source, at S, to a detector, at D (see Fig. 49). Let's put a screen in between the source and the detector and make two very tiny holes a few millimeters apart from each other, at A and B. (If the source and detector are 100 centimeters apart, the holes have to be smaller than a tenth of a millimeter.) Let's put A in line with S and D, and put B somewhere to the side of A, not in line with S and D.

When we close the hole at B, we get a certain number of clicks at D—which represents the photons that came through A (let's say the detector clicks an average of one time for every 100 photons that leave S, or 1%). When we close the hole at A and open the hole at B, we know from the second lecture that we get nearly the same number of clicks, on average, because the holes are so small. (When we "squeeze" light too much, the rules of the ordinary world—such as light goes in straight lines—fall apart.)

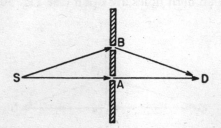

FIGURE 49.   *Two tiny holes (at A and B) in a screen that is between a source S and a detector D let nearly the same amount of light through (in this case 1%) when one or the other hole is open. When both holes are open, "interference" occurs: the detector clicks from zero to 4% of the time, depending on the separation of A and B—shown in Figure 51 (a).*

When we open both holes we get a complicated answer, because interference is present: If the holes are a certain distance apart, we get more clicks than the expected 2% (the maximum is about 4%); if the two holes are a slightly different distance apart, we get no clicks at all.

One would normally think that opening a second hole would *always* increase the amount of light reaching the detector, but that's not what actually happens. And so saying that the light goes "either one way or the other" is false. I still catch myself saying, "Well, it goes either this way or that way," but when I say that, I have to keep in mind that

I mean in the sense of adding amplitudes: the photon has an amplitude to go one way, *and* an amplitude to go the other way. If the amplitudes oppose each other, the light won't get there—even though, in this case, both holes are open.

Now, here's an extra twist to the strangeness of Nature that I'd like to tell you about. Suppose we put in some special detectors—one at A and one at B (it is possible to design a detector that can tell whether a photon went through it)—so we can tell through which hole(s) the photon goes when both holes are open (see Fig. 50). Since the

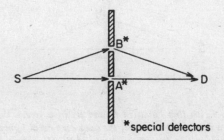

*special detectors

FIGURE 50. *When special detectors are put in at A and B to tell which way the light went when both holes are open, the experiment has been changed. Because a photon always goes through one hole or the other (when you are checking the holes), there are two distinguishable final conditions: 1) the detectors at A and D go off, and 2) the detectors at B and D go off. The probability of either event happening is about 1%. The probabilities of the two events are added in the normal way, which accounts for a 2% probability that the detector at D goes off—shown in Figure 51(b).*

probability that a single photon will get from S to D is affected only by the distance between the holes, there must be some sneaky way that the photon divides in two and then comes back together again, right? According to this hypothesis, the detectors at A and B should always go off together (at half strength, perhaps?), while the detector at D should go off with a probability of from zero to 4%, depending on the distance between A and B.

Here's what actually happens: the detectors at A and B *never* go off together—either A *or* B goes off. The photon does not divide in two; it goes one way or the other.

Furthermore, under such conditions the detector at D goes off 2% of the time—the simple sum of the probabilities for A and B (1% + 1%). The 2% is not affected by the spacing between A and B; the interference *disappears* when detectors are put in at A and B!

Nature has got it cooked up so we'll never be able to figure out how She does it: if we put instruments in to find out which way the light goes, we can find out, all right, but the wonderful interference effects disappear. But if we don't have instruments that can tell which way the light goes, the interference effects come back! Very strange, indeed!

To understand this paradox, let me remind you of a most important principle: in order to correctly calculate the probability of an event, one must be very careful to *define the complete event clearly*—in particular, what the initial conditions and the final conditions of the experiment are. You look at the equipment before and after the experiment, and look for changes. When we were calculating the probability that a photon gets from S to D with no detectors at A or B, the event was, simply, the detector at D makes a click. When a click at D was the only change in conditions, there was no way to tell which way the photon went, so there was interference.

When we put in detectors at A and B, we changed the problem. Now, it turns out, there are *two* complete events—two sets of final conditions—that are distinguishable: 1) the detectors at A and D go off, or 2) the detectors at B and D go off. When there are a number of possible final conditions in an experiment, we must calculate the probability of each as a separate, complete event.

To calculate the amplitude that the detectors at A and D go off, we multiply the arrows that represent the following steps: a photon goes from S to A, the photon goes from A to D, and the detector at D goes off. The square of the final arrow is the probability of this event—1%—the same as when the hole at B was closed, because both cases have exactly the same steps. The other complete event is the detectors at B and D go off. The probability of this event is calculated in a similar way, and is also the same as before—about 1%.

If we want to know how often the detector at D goes off and we don't care whether it was A or B that went off in the process, the probability is the simple sum of the two events—2%. In principle, if there is something left in the system that we *could have* observed to tell which way the photon went, we have different "final states" (distinguishable final conditions), and we add the *probabilities*—not the amplitudes—for each final state.[1]

I have pointed out these things because the more you see how strangely Nature behaves, the harder it is to make a model that explains how even the simplest phenomena actually work. So theoretical physics has given up on that.

We saw in the first lecture how an event can be divided into alternative ways and how the arrow for each way can be "added." In the second lecture, we saw how each way can be divided into successive steps, how the arrow for each step can be regarded as the transformation of a unit arrow,

[1] The complete story on this situation is very interesting: if the detectors at A and B are not perfect, and detect photons only *some* of the time, then there are *three* distinguishable final conditions: 1) the detectors at A and D go off; 2) the detectors at B and D go off, and 3) the detector at D goes off alone, with A and B unchanged (they are left in their initial state). The probabilities for the first two events are calculated in the way explained above (except that there will be an extra step—a shrink for the probability that the detector at A [or B] goes off, since the detectors are not perfect). When D goes off alone, we can't separate the two cases, and Nature plays with us by bringing in interference—the same peculiar answer we would have had if there were no detectors (except that the final arrow is shrunk by the amplitude that the detectors do *not* go off). The final result is a mixture, the simple sum of all three cases (see Fig. 51). As the reliability of the detectors increases, we get less interference.

and how the arrows for each step can be "multiplied" by successive shrinks and turns. We are thus familiar with all the necessary rules for drawing and combining arrows (that represent bits and pieces of events) to obtain a final arrow, whose square is the probability of an observed physical event.

It is natural to wonder how far we can push this process of splitting events into simpler and simpler subevents. What are the smallest possible bits and pieces of events? Is there

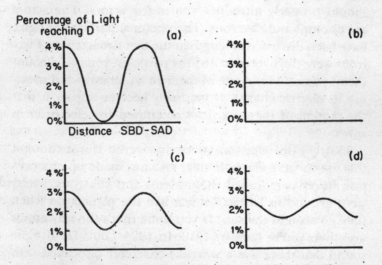

FIGURE 51.   *When there are no detectors at A or B, there is interference—
the amount of light varies from zero to 4% (a). When there are detectors at
A and B that are 100% reliable, there is no interference—the amount of light
reaching D is a constant 2% (b). When the detectors at A and B are not 100%
reliable (i.e., when sometimes there is nothing left in A or in B that can be
detected), there are now three possible final conditions—A and D go off, B
and D go off, and D goes off alone. The final curve is thus a mixture, made
up of contributions from each possible final condition. When the detectors at
A and B are less reliable, there is more interference present. Thus the detectors
in case (c) are less reliable than in case (d). The principle regarding interference
is: The probability of each of the different possible final conditions must be
independently calculated by adding arrows and squaring the length of the
final arrow; after that, the several probabilities are added together in the normal
fashion.*

a limited number of bits and pieces that can be compounded to form *all* the phenomena that involve light and electrons? Is there a limited number of "letters" in this language of quantum electrodynamics that can be combined to form "words" and "phrases" that describe nearly every phenomenon of Nature?

The answer is yes; the number is three. There are only three basic actions needed to produce all of the phenomena associated with light and electrons.

Before I tell you what these three basic actions are, I should properly introduce you to the actors. The actors are photons and electrons. The photons, particles of light, have been discussed at length in the first two lectures. Electrons were discovered in 1895 as particles: you could count them; you could put one of them on an oil drop and measure its electric charge. It gradually became apparent that the motion of these particles accounted for electricity in wires.

Shortly after electrons were discovered it was thought that atoms were like little solar systems, made up of a central, heavy part (called the nucleus) and electrons, which went around in "orbits," much like the planets do when they go around the sun. If you think that's the way atoms are, then you're back in 1910. In 1924 Louis De Broglie found that there was a wavelike character associated with electrons, and soon afterwards, C. J. Davisson and L. H. Germer of the Bell Laboratories bombarded a nickel crystal with electrons and showed that they, too, bounced off at crazy angles (just like X-rays do), and that these angles could be calculated from De Broglie's formula for the wavelength of an electron.

When we look at photons on a large scale—much larger than the distance required for one stopwatch turn—the phenomena that we see are very well approximated by rules such as "light travels in straight lines," because there are enough paths around the path of minimum time to reinforce each other, and enough other paths to cancel each other out. But when the space through which a photon moves becomes too small (such as the tiny holes in the screen), these rules fail—we discover that light doesn't have to go in straight lines, there are interferences created by two holes, and so on. The same situation exists with electrons: when seen on a large scale, they travel like particles, on definite paths. But on a small scale, such as inside an atom, the space is so small that there is no main path, no "orbit"; there are all sorts of ways the electron could go, each with an amplitude. The phenomenon of interference becomes very important, and we have to sum the arrows to predict where an electron is likely to be.

It's rather interesting to note that electrons looked like particles at first, and their wavish character was later discovered. On the other hand, apart from Newton making a mistake and thinking that light was "corpuscular," light looked like waves at first, and its characteristics as a particle were discovered later. In fact, both objects behave somewhat like waves, and somewhat like particles. In order to save ourselves from inventing new words such as "wavicles," we have chosen to call these objects "particles," but we all know that they obey these rules for drawing and combining arrows that I have been explaining. It appears that *all* the "particles" in Nature—quarks, gluons, neutrinos, and so forth (which will be discussed in the next lecture)—behave in this quantum mechanical way.

So now, I present to you the three basic actions, from which all the phenomena of light and electrons arise.

—ACTION #1: A photon goes from place to place.
—ACTION #2: An electron goes from place to place.
—ACTION #3: An electron emits or absorbs a photon.

Each of these actions has an amplitude—an arrow—that can be calculated according to certain rules. In a moment, I'll tell you those rules, or laws, out of which we can make the whole world (aside from the nuclei, and gravitation, as always!).

Now, the stage on which these actions take place is not just space, it is space and time. Until now, I have disregarded problems concerning time, such as exactly when a photon leaves the source and exactly when it arrives at the detector. Although space is really three-dimensional, I'm going to reduce it to one dimension on the graphs that I'm going to draw: I will show a particular object's location in space on the horizontal axis, and the time on the vertical axis.

The first event I am going to draw in space and time— or space-time, as I might inadvertently call it—is a baseball standing still (See Fig. 52). On Thursday morning, which

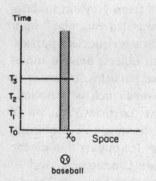

FIGURE 52. *The stage on which all actions in the universe take place is space-time. Usually consisting of four dimensions (three for space and one for time), space-time will be represented here in two dimensions—one for space, in the horizontal dimension, and one for time, in the vertical. Each time we look at the baseball (such as at time T_3), it is in the same place. This produces a "band of baseball" going straight up, as time goes on.*

I will label as T_0, the baseball occupies a certain space, which I will label as X_0. A few moments later, at T_1, it occupies the same space, because it's standing still. A few moments later, at T_2, the baseball is still at X_0. So the diagram of a baseball standing still is a vertical band, going straight up, with baseball all over it inside.

What happens if we have a baseball drifting in the weightlessness of outer space, going straight toward a wall? Well, on Thursday morning (T_0) it starts at X_0 (see Fig. 53), but a little bit later, it's not in the same place—it has drifted over a little bit, to X_1. As the baseball continues to drift, it creates a slanted "band of baseball" on the diagram of space-time. When the baseball hits the wall (which is standing still and is therefore a vertical band), it goes back the other way, exactly where it came from in space (X_0), but to a different point in time (T_6).

FIGURE 53. *A baseball drifting directly toward a wall at right angles and then bouncing back to its original location (shown below the graph) is moving in one dimension and appears as a slanted "band of baseball." At times T_1 and T_2, the baseball is getting closer to the wall; at T_3 it hits the wall, and begins to go back.*

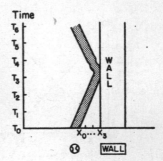

As for the time scale, it is most convenient to represent the time not in seconds, but in much smaller units. Since we will be dealing with photons and electrons, which move very rapidly, I am going to have a 45° angle represent something going the speed of light. For example, for a particle moving at the speed of light from X_1T_1 to X_2T_2, the horizontal distance between X_1 and X_2 is the same as the vertical distance between T_1 and T_2 (see Fig. 54). The factor by which time is stretched out (to make a 45° angle represent a particle going the speed of light) is called c, and you'll find c's flying around everywhere in Einstein's formulas—they are the result of the unfortunate choice of the second as the unit of time, rather than the time it takes light to go one meter.

Now, let's look at the first basic action in detail—a photon goes from place to place. I will draw this action as a wiggly line from A to B for no good reason. I should be more careful: I should say, a photon that is known to be at a given place at a given time has a certain amplitude to get to another place at another time. On my space-time graph (see Fig. 55), the photon at point A—at X_1 and T_1—has an amplitude to appear at point B—X_2 and T_2. The size of this amplitude I will call P(A to B).

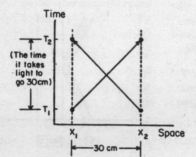

FIGURE 54. *The time scale I will use in these graphs will show particles going at the speed of light to be travelling at a 45-degree angle through space-time. The amount of time it takes light to go 30 centimeters—from X_1 to X_2 or from X_2 to X_1—is about one-billionth of a second.*

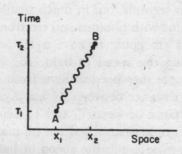

FIGURE 55. *A photon (represented by a wavy line) has an amplitude to go from a point A in space-time to another point, B. This amplitude, which I will call P(A to B), is calculated from a formula that depends only on the difference in location—$(X_2 - X_1)$—and the difference of the time—$(T_2 - T_1)$. In fact, it's a simple function that is the inverse of the difference of their squares—an "interval," I, that can be written as $(X_2 - X_1)^2 - (T_2 - T_1)^2$.*

There is a formula for the size of this arrow, P(A to B). It is one of the great laws of Nature, and it's very simple. It depends on the difference in *distance* and the difference in *time* between the two points. These differences can be expressed mathematically[2] as $(X_2 - X_1)$ and $(T_2 - T_1)$.

The major contribution to P(A to B) occurs at the conventional speed of light—when $(X_2 - X_1)$ is equal to $(T_2 - T_1)$—where one would expect it all to occur, but there is also an amplitude for light to go faster (or slower) than the conventional speed of light. You found out that in the last lecture that light doesn't go only in straight lines; now, you find out that it doesn't go only at the speed of light!

It may surprise you that there is an amplitude for a photon to go at speeds faster or slower than the conventional speed, c. The amplitudes for these possibilities are very small compared to the contribution from speed c; in fact, they cancel out when light travels over long distances. However, when the distances are short—as in many of the

---

[2] In these lectures, I am plotting a point's location in space in one dimension, along the x-axis. To locate a point in three-dimensional space, a "room" has to be set up, and the distance of the point from the floor and from each of two adjacent walls (all at right angles to each other) has to be measured. These three measurements can be labeled X_1, Y_1, and Z_1. The actual distance from this point to a second point with measurements X_2, Y_2, Z_2 can be calculated using a "three-dimensional Pythagorean Theorem": the square of this actual distance is

$$(X_2 - X_1)^2 + (Y_2 - Y_1)^2 + (Z_2 - Z_1)^2.$$

The excess of *this* over the time difference, squared—

$$(X_2 - X_1)^2 + (Y_2 - Y_1)^2 + (Z_2 - Z_1)^2 - (T_2 - T_1)^2$$

—is sometimes called "the Interval," or I, and is the combination that Einstein's theory of relativity says that P(A to B) must depend on. Most of the contribution to the final arrow for P(A to B) is just where you would expect it—where the difference in distance is equal to the difference in time (that is, when I is zero). But in addition; there is a contribution when I is not zero, that is inversely proportional to I: it points in the direction of 3 o'clock when I is more than zero (when light is going faster than c), and points toward 9 o'clock when I is less than zero. These later contributions cancel out in many circumstances (see Fig. 56).

diagrams I will be drawing—these other possibilities become vitally important and must be considered.

So that's the first basic action, the first basic law of physics—a photon goes from point to point. That explains all about optics; that's the entire theory of light! Well, not quite: I left out polarization (as always), and the interaction of light with matter, which brings me to the second law.

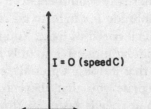

I = O (speed C)

(slower than C) I < O   I > O (faster than C)

FIGURE 56. *When light goes at the speed* C, *the "interval,"* I, *equals zero, and there is a large contribution in the 12 o'clock direction. When* I *is greater than zero, there is a small contribution in the three o'clock direction inversely proportional to* I; *when* I *is less than zero, there is a similar contribution in the nine o'clock direction. Thus light has an amplitude to go faster or slower than speed* C, *but these amplitudes cancel out over long distances.*

The second action fundamental to quantum electrodynamics is: An electron goes from point A to point B in space-time. (For the moment we will imagine this electron as a simplified, fake electron, with no polarization—what the physicists call a "spin-zero" electron. In reality, electrons have a type of polarization, which doesn't add anything to the main ideas; it only complicates the formulas a little bit.) The formula for the amplitude for this action, which I will call E(A to B) also depends on $(X_2 - X_1)$ and $(T_2 - T_1)$ (in the same combination as described in note 2) as well as on a number I will call "n," a number that, once determined, enables all our calculations to agree with experiment. (We will see later how we determine n's value.) It is a rather complicated formula, and I'm sorry that I don't know how to explain it in simple terms. However, you might be interested to know that the formula for P(A to B)—a photon going from place to place in space-time—is the same as that for E(A to B)—an electron going from place to place—if n is set to zero.[3]

The third basic action is: an electron emits or absorbs a photon—it doesn't make any difference which. I will call this action a "junction," or "coupling." To distinguish electrons from photons in my diagrams, I will draw each electron going through space-time as a straight line. Every coupling, therefore, is a junction between two straight lines and a wavy line (see Fig. 58). There is no complicated formula for the amplitude of an electron to emit or absorb a photon; it doesn't depend on anything—it's just a number! This junction number I will call j—its value is about -0.1: a shrink to about one-tenth, and half a turn.[4]

Well, that's all there is to these basic actions—except for some slight complications due to this polarization that we're

[3] The formula for E(A to B) is complicated, but there is an interesting way to explain what it amounts to. E(A to B) can be represented as a giant sum of a lot of different ways an electron could go from point A to point B in space-time (see Fig. 57): the electron could take a "one-hop flight," going directly from A to B; it could take a "two-hop flight," stopping at an intermediate point C; it could take a "three-hop flight," stopping at points D and E, and so on. In such an analysis, the amplitude for each "hop"—from one point F to another point G—is P(F to G), the same as the amplitude for a photon to go from a point F to a point G. The amplitude for each "stop" is represented by n^2, n being the same number I mentioned before which we used to make our calculations come out right.

The formula for E(A to B) is thus a series of terms: P(A to B) [the "one-hop" flight] + P(A to C)$*n^2*$P(C to B) ["two-hop" flights, stopping at C] + P(A to D)$*n^2*$P(D to E) $* n^2*$P(E to B) ["three-hop" flights, stopping at D and E] + ... for *all possible intermediate points* C, D, E, and so on.

Note that when n increases, the nondirect paths make a greater contribution to the final arrow. When n is zero (as for the photon), all terms with an n drop out (because they are also equal to zero), leaving only the first term, which is P(A to B). Thus E(A to B) and P(A to B) are closely related.

[4] This number, the amplitude to emit or absorb a photon, is sometimes called the "charge" of a particle.

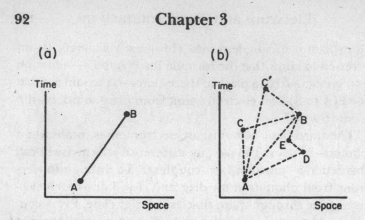

(a)   (b)

FIGURE 57. *An electron has an amplitude to go from point to point in space-time, which I will call "E(A to B)." Although I will represent E(A to B) as a straight line between two points (a), we can think of it as the sum of many amplitudes (b)—among them, the amplitude for the electron to change direction at points C or C' on a "two-hop" path, and the amplitude to change direction at D and E on a "three-hop" path—in addition to the direct path from A to B. The number of times an electron can change direction is anywhere from zero to infinity, and the points at which the electron can change direction on its way from A to B in space-time are infinite. All are included in E(A to B).*

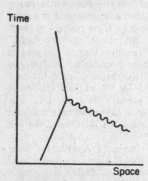

FIGURE 58. *An electron, depicted by a straight line, has a certain amplitude to emit or absorb a photon, shown by a wavy line. Since the amplitude to emit or absorb is the same, I will call either case a "coupling." The amplitude for a coupling is a number that I will call j; it is about −0.1 for the electron (this number is sometimes called the "charge").*

always leaving out. Our next job is to put these three actions together to represent circumstances that are somewhat more complicated.

For our first example, let's calculate the probability that two electrons, at points 1 and 2 in space-time, end up at points 3 and 4 (see Fig. 59). This event can happen in several ways. The first way is that the electron at 1 goes to 3—computed by putting 1 and 3 into the formula E(A to B), which I will write as E(1 to 3)—and the electron at 2 goes to 4—computed by E(2 to 4). These are two "subevents" happening concomitantly, so the two arrows are multiplied to produce an arrow for this first way the event could happen. Therefore we write the formula for the "first-way arrow" as E(1 to 3) * E(2 to 4).

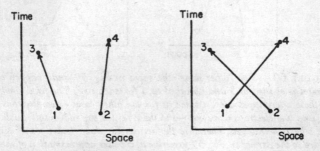

FIGURE 59.   *To calculate the probability that electrons at points 1 and 2 in space-time end up at points 3 and 4, we calculate the "first way" arrow for 1 going to 3 and 2 going to 4 with the formula for E(A to B); then we calculate the "second way" arrow for 1 going to 4 and 2 going to 3 (a "crossover"). Finally, we add the "first way" and "second way" arrows to arrive at a good approximation of the final arrow. (This is true for the fake, simplified "spin zero" electron. Had we included the polarization of the electron, we would have subtracted—rather than added—the two arrows.)*

Another way this event could happen is that the electron at 1 goes to 4 and the electron at 2 goes to 3—again, two concomitant subevents. The "second-way arrow" is E(1 to 4) * E(2 to 3), and we add it to the "first-way" arrow.[5]

This is a good approximation for the amplitude of this event. To make a more exact calculation that will agree more closely with the results of experiment, we must con-

---

[5] Had I included the effects of the polarization of the electron, the "second-way" arrow would have been "subtracted"—turned 180° and added. (More on this comes later in this lecture.)

sider other ways this event could happen. For instance, for
each of the two main ways the event can happen, one electron could go charging off to some new and wonderful
place and emit a photon (see Fig. 60). Meanwhile, the other
electron could go to some other place and absorb the pho-

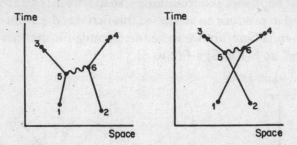

FIGURE 60. *Two "other ways" the event in Fig. 59 could happen are: a
photon is emitted at 5 and absorbed at 6 for each case. The final conditions
of these alternatives are the same as for the other cases—two electrons went
in, and two electrons came out—and these results are indistinguishable from
the other alternatives. Therefore the arrows for these "other ways" must be
added to the arrows in Fig. 59 to arrive at a better approximation of the final
arrow for the event.*

ton. Calculating the amplitude for the first of these new
ways involves multiplying the amplitudes for: an electron
goes from 1 to the new and wonderful place, 5 (where it
emits a photon), and then goes from 5 to 3; the other
electron goes from 2 to the other place, 6 (where it absorbs
the photon), and then goes from 6 to 4. We must remember
to include the amplitude that the photon goes from 5 to
6. I'm going to write the amplitude for this way the event
could happen in a high-class mathematical fashion, and you
can follow along: E(1 to 5)*j*E(5 to 3) * E(2 to 6)*j*E(6 to
4) * P(5 to 6)—a lot of shrinking and turning. (I'll let you
figure out the notation for the other case, where the electron at 1 ends up at 4, and the electron at 2 ends up at 3.)[6]

[6] The final conditions of the experiment for these more complicated
ways are the same as for the simpler ways—electrons start at points 1 and

But wait: positions 5 and 6 could be anywhere in space and time—yes, anywhere—and the arrows for *all* of those positions have to be calculated and added together. You see it's getting to be a lot of work. Not that the rules are so difficult—it's like playing checkers: the rules are simple, but you use them over and over. So our difficulty in calculating comes from having to pile so many arrows together. That's why it takes four years of graduate work for the students to learn how to do this efficiently—and we're looking at an *easy* problem! (When the problems get too difficult, we just put them on the computer!)

I would like to point out something about photons being emitted and absorbed: if point 6 is later than point 5, we might say that the photon was emitted at 5 and absorbed at 6 (see Fig. 61). If 6 is earlier than 5, we might prefer to say the photon was emitted at 6 and absorbed at 5, but we could just as well say that the photon is going backwards in time! However, we don't have to worry about which way in space-time the photon went; it's all included in the formula for P(5 to 6), and we say a photon was "exchanged." Isn't it beautiful how simple Nature is![7]

Now, in addition to the photon that is exchanged between 5 and 6, another photon could be exchanged—between two points, 7 and 8 (see Fig. 62). I'm too tired to write down all the basic actions whose arrows have to be multiplied, but—as you may have noticed—every straight line gets an E(A to B), every wavy line gets a P(A to B), and every coupling gets a *j*. Thus, there are six E(A to B)'s, two P(A to B)'s, and four *j*'s—for *every possible 5, 6, 7, and 8*! That makes billions of tiny arrows that have to be multiplied and then added together!

---

2 and end up at points 3 and 4—so we cannot distinguish between these alternatives and the first two. Therefore we must add the arrows for these two ways to the two ways just previously considered.

[7] Such an exchanged photon that never really appears in the initial or final conditions of the experiment is sometimes called a "virtual photon."

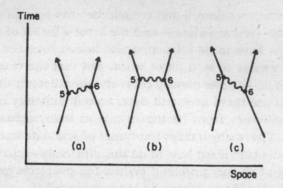

FIGURE 61. *Since light has an amplitude to go faster or slower than the conventional speed of light, the photons in all three examples above can be thought of as being emitted from point 5 and absorbed at point 6, even though the photon in example (b) is emitted at the same time that it is absorbed, and the photon in (c) is emitted later than it is absorbed—a situation in which you might have preferred to say that it was emitted by 6 and absorbed by 5; otherwise, the photon would have to go backwards in time! As far as calculating (and Nature) is concerned, it's all the same (and it's all possible), so we simply say a photon is "exchanged" and plug the locations in space-time into the formula for P(A to B).*

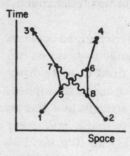

FIGURE 62. *Yet another way the event in Fig. 59 could happen is that two photons could be exchanged. Many diagrams of this way are possible (as we will see in more detail later); one of them is shown here. The arrow for this way involves all possible intermediate points 5, 6, 7, and 8, and is calculated with great difficulty. Because j is less than 0.1, the length of this arrow is generally less than 1 part in 10,000 (because there are four couplings involved) compared to the "first way" and "second way" arrows in Fig. 59 that contained no j's.*

It appears that calculating the amplitude for this simple event is a hopeless business, but when you're a graduate student you've got to get your degree, so you keep on going.

But there *is* hope for success. It is found in that magic number, *j*. The first two ways the event could happen had no j's in the calculation; the next way had $j*j$, and the last way we looked at had $j*j*j*j$. Since $j*j$ is less than 0.01, it means the length of the arrow for this way is generally less than 1% of the arrow for the first two ways; an arrow with $j*j*j*j$ in it is less than 1% of 1%—one part in 10,000—compared to the arrows that have no j. If you've got enough time on the computer, you can work out the possibilities that involve j^6—one part in a million—and match the accuracy of the experiments. That's how the calculations of simple events are made. That's the way it works; that's all there is to it!

Let's look at another event now. We begin with a photon and an electron, and we end with a photon and an electron. One way this event can happen is: a photon is absorbed by an electron, the electron continues on a bit, and a new photon comes out. This process is called the scattering of light. When we make the diagrams and calculations for scattering, we must include some peculiar possibilities (see Fig. 63). For example, the electron could emit a photon *before* absorbing one (b). Even more strange is the possibility

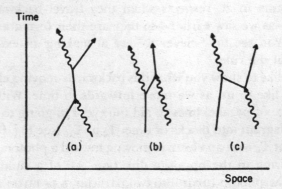

FIGURE 63. *The scattering of light involves a photon going into an electron and a photon coming out—not necessarily in that order, as seen in example (b). The example in (c) shows a strange but real possibility: the electron emits a photon, rushes backwards in time to absorb a photon, and then continues forwards in time.*

(c) that the electron emits a photon, then *travels backwards in time* to absorb a photon, and then proceeds forwards in time again. The path of such a "backwards-moving" electron can be so long as to appear real in an actual physical experiment in the laboratory. Its behavior is included in these diagrams and the equation for E(A to B).

The backwards-moving electron when viewed with time moving forwards appears the same as an ordinary electron, except it's attracted to normal electrons—we say it has a "positive charge." (Had I included the effects of polarization, it would be apparent why the sign of *j* for the backwards-moving electron appears reversed, making the charge appear positive.) For this reason it's called a "positron." The positron is a sister particle to the electron, and is an example of an "anti-particle."[8]

This phenomenon is general. Every particle in Nature has an amplitude to move backwards in time, and therefore has an anti-particle. When a particle and its anti-particle collide, they annihilate each other and form other particles. (For positrons and electrons annihilating, it is usually a photon or two.) And what about photons? Photons look exactly the same in all respects when they travel backwards in time—as we saw earlier—so they are their own anti-particles. You see how clever we are at making an exception part of the rule!

I'd like to show you what this backwards-moving electron looks like to us, as we move forwards in time. With a sequence of parallel lines to aid the eye, I'm going to divide the diagram into blocks of time, T_0 to T_{10} (see Fig. 64). We start at T_0 with an electron moving toward a photon, which is moving in the opposite direction. All of a sudden—at T_3—the photon turns into two particles, a positron and an

---

[8] Dirac proposed the reality of "anti-electrons" in 1931; in the following year, Carl Anderson found them experimentally and called them "positrons." Today, positrons can be easily made (for example, by making two photons collide with each other) and kept for weeks in a magnetic field.

FIGURE 64. *Looking at example (c) from Fig. 63 going only forwards in time (as we are forced to do in the laboratory), from T_0 to T_3 we see the electron and photon moving toward each other. All of a sudden, at T_3 the photon "disintegrates" and two particles appear—an electron and a new kind of particle (called a "positron") which is an electron going backwards in time and which appears to move toward the original electron (itself!). At T_5 the positron annihilates with the original electron to produce a new photon. Meanwhile, the electron created by the earlier photon continues forwards in space-time. This sequence of events has been observed in the laboratory, and is included automatically in the formula for E(A to B) without any modification.*

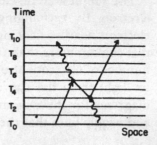

electron. The positron doesn't last very long: it soon runs into the electron—at T_5, where they annihilate and produce a new photon. Meanwhile, the electron created earlier by the original photon continues on through space-time.

The next thing I would like to talk about is an electron in an atom. In order to understand the behavior of electrons in atoms, we have to add one other feature, the nucleus—the heavy part at the center of an atom that contains at least one proton (a proton is a "Pandora's Box" that we will open in the next lecture). I will not give you the correct laws for the behavior of the nucleus in this lecture; they are very complicated. But in this case, where the nucleus is quiet, we can approximate its behavior as that of a particle with an amplitude to go from one place to another in space-time according to the formula for E(A to B), but with a much higher number for n. Since the nucleus is so heavy compared to an electron, we can deal with it approximately here by saying that it stays in essentially one place as it moves through time.

The simplest atom, called hydrogen, is a proton and an electron. By exchanging photons, the proton keeps the electron nearby, dancing around it (see Fig. 65).[9] Atoms that contain more than one proton and the corresponding number of electrons also scatter light (atoms in the air scatter light from the sun and make the sky blue), but the diagrams for these atoms would involve so many straight and wiggly lines that they'd be a complete mess!

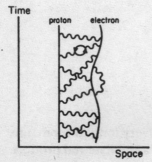

FIGURE 65. *An electron is kept within a certain range of distance to the nucleus of an atom by photon exchanges with a proton (a "Pandora's Box" that we will look into in Chapter 4). For now, the proton can be approximated as a stationary particle. Shown here is a hydrogen atom, consisting of a proton and an electron exchanging photons.*

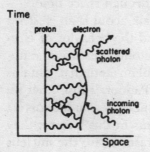

FIGURE 66. *The scattering of light by an electron in an atom is the phenomenon that accounts for partial reflection in a layer of glass. The diagram shows one way this event can happen in a hydrogen atom.*

Now, I'd like to show you a diagram of an electron in a hydrogen atom scattering light (see Fig. 66). As the electron and the nucleus are exchanging photons, a photon comes

[9] The amplitude for the photon exchange is $(-j) * P(A - B) * j$—two couplings and the amplitude for a photon to go from place to place. The amplitude for a proton to have a coupling with a photon is $-j$.

from outside the atom, hits the electron and is absorbed; then a new photon is emitted. (As usual, there are other possibilities to be considered, such as the new photon is emitted before the old photon is absorbed.) The total amplitude for all the ways an electron can scatter a photon can be summed up as a single arrow, a certain amount of shrink and turn. (Later, we will call this arrow "S.") This amount depends on the nucleus and the arrangement of the electrons in the atoms, and is different for different materials.

Now, let's look again at the partial reflection of light by a layer of glass. How does it work? I talked about light being reflected from the front surface and the back surface. This idea of surfaces was a simplification I made in order to keep things easy at the beginning. Light is really not affected by surfaces. An incoming photon is scattered by the electrons in the atoms inside the glass, and a *new* photon comes back up to the detector. It's interesting that instead of adding up all the billions of tiny arrows that represent the amplitude for all the electrons inside the glass to scatter an incoming photon, we can add just two arrows—for the "front surface" and "back surface" reflections—and come out with the same answer. Let's see why.

To discuss reflection by a layer from our new point of view we must take into account the dimension of time. Previously, when we talked about light from a monochromatic source, we used an imaginary stopwatch that times a photon as it moves—the hand of this stopwatch determined the angle of the amplitude for a given path. In the formula for P(A to B) (the amplitude for a photon to go from point to point) there is no mention of any turning. What happened to the stopwatch? What happened to the turning?

In the first lecture I simply said that the light source was monochromatic. To correctly analyze partial reflection by a layer, we need to know more about a monochromatic

light source. The amplitude for a photon to be emitted by
a source varies, in general, with the *time*: as time goes on,
the angle of the amplitude for a photon to be emitted by
a source changes. A source of white light—many colors
mixed together—emits photons in a chaotic manner: the
angle of the amplitude changes abruptly and irregularly in
fits and starts. But when we construct a *monochromatic*
source, we are making a device that has been carefully
arranged so that the amplitude for a photon to be emitted
at a certain time is easily calculated: it changes its angle at
a *constant* speed, like a stopwatch hand. (Actually, this arrow
turns at the same speed as the imaginary stopwatch we used
before, but in the opposite direction—see Fig. 67.)

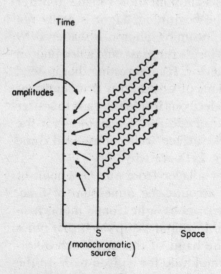

FIGURE 67. *A mono-
chromatic source is a beau-
tifully constructed apparatus
that emits a photon in a very
predictable way: the amplitude for a photon to be emitted at a certain time rotates
counterclockwise as time
moves forwards. Thus the
amplitude for the source to
emit a photon at a later time
has a lesser angle. It will be
assumed that all the light
emitted from the source goes
at speed c (since the distances
are large).*

The rate of turning depends on the color of the light: the
amplitude for a blue source turns nearly twice as fast as
that for a red source, just as before. So the timer we used
for the "imaginary stopwatch" was the monochromatic
source:—in reality, the angle of the amplitude for a given

path depends on what *time* the photon is emitted from the source.

Once a photon has been emitted, there is no further turning of the arrow as a photon goes from one point to another in space-time. Although the formula P(A to B) says that there is an amplitude for light to go from one place to another at speeds *other* than *c*, the distance from the source to the detector in our experiment is relatively large (compared to an atom), so the only surviving contribution to P(A to B)'s length that counts comes from speed *c*.

To begin our new calculation of partial reflection, let's start by defining the event completely: the detector at A makes a click at a certain *time*, T. Then, let's divide the layer of glass into a number of very thin sections—let's say, six (see Fig. 68a). From the analysis we did in the second lecture in which we found that nearly all the light is reflected from the middle of a mirror, we know that although each electron is scattering light in all directions, when all the arrows for each section are added, the only place where they *don't* cancel out is where light goes straight down to the middle of the section and scatters in one of two directions—straight back up to the detector or straight down through the glass. The final arrow for the event will thus be determined by adding the six arrows representing the scattering of light from the six middle points—X_1 to X_6—arranged vertically throughout the glass.

All right, let's calculate the arrow for each of these ways the light could go—via the six points, X_1 to X_6. There are four steps involved in each way (which means four arrows will be multiplied):

—STEP #1: A photon is emitted by the source at a certain time.
—STEP #2: The photon goes from the source to one of the points in the glass.

(a)

(b)

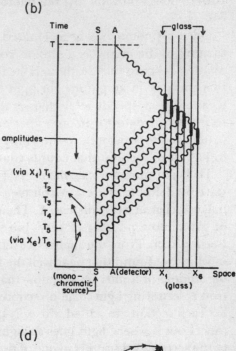

(c)

(d)

FIGURE 68. *We begin our new analysis of partial reflection by dividing a layer of glass into a number of sections (here, six), and looking at the various ways the light could go from the source to the glass and back up to the detector at A. The only important points in the glass (where the amplitudes for scattering light don't cancel out) are located at the middle of each section; X_1 to X_6 are shown in (a) at their physical location inside the glass, and in (b) as vertical lines on the space-time graph. The event whose probability we are calculating is: the detector at A makes a click at a certain time, T. Thus the event appears as a point (where A and T intersect) on the space-time graph.*

For each of the ways the event can happen, four steps must occur in succession, so four arrows have to be multiplied. The steps are shown in (b): 1) a photon leaves the source at a certain time (the arrows at T_1 to T_6 represent *the amplitude to do that for six different times); 2) the photon goes from the source to one of the points in the glass (the six alternatives are depicted as wavy lines going up to the right); 3) an electron at one of the points scatters a photon (shown as short, wide vertical lines); and 4) a new photon goes to the detector and arrives at the appointed time, T (shown as a wavy line going up to the left). The amplitudes for steps 2, 3, and 4 are the same for the six alternatives, while the amplitudes for step 1 are different: compared to a photon scattered by an electron at the top of the glass (at X_1), a photon scattered deeper in the glass—at X_2, for example—must leave the source earlier, at T_2.*

When we are finished multiplying the four arrows for each alternative, the resulting arrows, shown in (c), are shorter than those in (b); each has been turned 90° (in accordance with the scattering characteristics of electrons in glass). When these six arrows are added together in order, they form an arc; the final arrow is its chord. The same final arrow can be obtained by drawing two radius arrows, shown in (d), and "subtracting" them (turning the "front surface" arrow around in the opposite direction and adding it to the "back surface" arrow). This shortcut was used as a simplification in the first lecture.

—STEP #3: The photon is scattered by an electron at that point.

—STEP #4: A new photon makes its way up to the detector.

We will say the amplitudes for steps 2 and 4 (a photon goes to or from a point in the glass) involve no shrinking or turning, because we can assume that none of the light gets lost or spread out between the source and the glass or between the glass and the detector. For step 3 (an electron scatters a photon) the amplitude for scattering is a constant—a shrink and a turn by a certain amount, S—and is the same everywhere in the glass. (This amount is, as I mentioned before, different for different materials. For glass, the turn of S is 90°.) Therefore, of the four arrows to be multiplied, only the arrow for step 1—the amplitude for a photon to be emitted from the source at a certain time—is different from one alternative to the next.

The time at which a photon would have to have been emitted to reach the detector A at time T (see Fig. 68b) is not the same for the six different paths. A photon scatterd by X_2 would have to have been emitted slightly *earlier* than a photon scattered by X_1, because that path is longer. Thus the arrow at T_2 is turned slightly more than the arrow at T_1 because the amplitude for a monochromatic source to emit a photon at a certain time rotates counterclockwise as time goes on. The same goes for each arrow down to T_6: all six arrows have the same length, but they are turned at different angles—that is, they are pointing in different directions—because they represent a photon emitted by the source at different times.

After shrinking the arrow at T_1 by the amounts prescribed in steps 2, 3 and 4—and turning it the 90° prescribed in step 3—we end up with arrow 1 (see Fig. 68c). The same goes for the arrows 2 through 6. Thus arrows 1 through 6 are all the same (shortened) length, and are turned relative to each other in exactly the same amount as the arrows at T_1 through T_6.

Next, we add arrows 1 to 6. Connecting the arrows in order from 1 to 6, we get something like an arc, or part of a circle. The final arrow forms the chord of this arc. The length of the final arrow increases with the thickness of the glass—thicker glass means more sections, more arrows, and therefore more of a circle—until half a circle is reached (and the final arrow is its diameter). Then the length of the final arrow *decreases* as the thickness of the glass continues to increase, and the circle becomes complete to begin a new cycle. The square of this length is the probability of the event, and it varies in the cycle of zero to 16%.

There is a mathematical trick we can use to get the same answer (see Fig. 68d): If we draw arrows from the center of the "circle" to the tail of arrow 1 and to the head of arrow 6, we get two radii. If the radius arrow from the center to arrow 1 is turned 180° ("subtracted"), then it can be combined with the other radius arrow to give us the same final arrow! That's what I was doing in the first lecture: these two radii are the two arrows I said represented the "front surface" and "back surface" reflections. They each have the famous length of 0.2.[10]

Thus we can get the correct answer for the probability of partial reflection by imagining (falsely) that all reflection comes from only the front and back surfaces. In this intuitively easy analysis, the "front surface" and "back surface" arrows are mathematical constructions that give us the right answer, whereas the analysis we just did—with the space-time drawing and the arrows forming part of a circle—is a more accurate representation of what is really going on: partial reflection is the scattering of light by electrons *inside* the glass.

Now, what about the light that goes *through* the layer of glass? First, there is an amplitude that the photon goes straight through the glass without hitting any electrons (see Fig. 69a). This is the most important arrow in terms of length. But there are six other ways a photon could reach the detector below the glass: a photon could hit X_1 and scatter the new photon down to B; a photon could hit X_2 and scatter the new photon down to B, and so on. These six arrows all have the same length as the arrows that formed the "circle" in the previous example: their length

---

[10] The radius of the arc evidently depends on the *length* of the arrow for each section, which is ultimately determined by the amplitude S that an electron in an atom of glass scatters a photon. This radius can be calculated using the formulas for the three basic actions for the multitude of photon exchanges involved and summing up the amplitudes. It is a very difficult problem, but the radius has been calculated for relatively simple substances with considerable success, and the variation of the radius from substance to substance is fairly well understood using these ideas of quantum electrodynamics. It must be said, however, that no direct calculation from first principles for a substance as complex as glass has ever actually been done. In such cases, the radius is determined by experiment. For glass, it has been determined from experiment that the radius is approximately 0.2 (when the light shines directly onto the glass at right angles).

Chapter 3

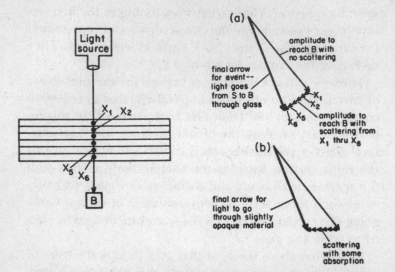

FIGURE 69. *The largest amplitude for light that is transmitted through the layer of glass to the detector at B comes from the part that represents no scattering by the electrons inside the glass, shown in (a). To this arrow we add six small arrows that represent the scattering of light from each of the sections, represented by points X_1 to X_6. These six arrows have the same length (because the amplitude for scattering is the same anywhere in the glass) and point in the same direction (because the length of each path from the source through any point X to B is the same). After adding the small arrows to the large one, we find the final arrow for the transmission of light through a layer of glass is turned more than what we would have expected if the light came only directly. For this reason it appears to us that light takes longer to go through glass than it takes to go through a vacuum or through air. The amount of turning by the final arrow caused by the electrons in a material is called the "index of refraction."*

For transparent materials, the little arrows are at right angles to the main arrow (they actually curve around when we include double and triple scatterings, keeping the final arrow from being longer than the main arrow: Nature always has it worked out so we never get more light out than we put in). For materials that are partially opaque—that absorb light to an extent—the little arrows point toward the main arrow, resulting in a final arrow that is significantly shorter than expected, shown in (b). This shorter final arrow represents a reduced probability of a photon being transmitted through partially opaque material.

is based on that same amplitude of an electron in the glass to scatter a photon, S. But this time, all six arrows point in the same direction, because the length of all six paths that involve one scattering is the same. The direction of these minor arrows is at right angles to the main arrow for transparent substances such as glass. When the minor arrows are added to the main arrow, they result in a final arrow that has the same length as the main arrow, but is turned in a slightly different direction. The thicker the glass, the more minor arrows there are, and the more the final arrow is turned. That's how a focusing lens really works: the final arrows for all the paths can be made to point in the same direction by inserting extra thicknesses of glass into the shorter paths.

The same effect would appear if photons went slower through glass than through air: there would be extra turning of the final arrow. That's why I said earlier that light appears to go slower through glass (or water) than through air. In reality, the "slowing" of the light is extra turning caused by the atoms in the glass (or water) scattering the light. The degree to which there is extra turning of the final arrow as light goes through a given material is called its "index of refraction."[11]

For substances that absorb light, the minor arrows are at

[11] Each of the arrows for reflection by a section (that form a "circle") has the same length as each of the arrows that make the final arrow from transmission appear to turn more. Thus there is a relationship between the partial reflection of a material and its index of refraction.

It appears that the final arrow has become longer than 1, which means that more light comes out through the glass than went into it! It looks that way because I disregarded the amplitudes for a photon to go down to one section, a new photon to scatter up to another section, and then a third photon to scatter back down through the glass—and other, more complicated possibilities—which result in the little arrows curving around and keeping the length of the final arrow between 0.92 and 1 (so the total probability of light being reflected or transmitted by the layer of glass is always 100%).

less than right angles to the main arrow (see Fig. 69b). This causes the final arrow to be shorter than the main arrow, indicating that the probability of a photon going through partially opaque glass is smaller than through transparent glass.

Thus it is that all the phenomena and the arbitrary numbers mentioned in the first two lectures—such as partial reflection with an amplitude of 0.2, the "slowing" of light in water and glass, and so on—are explained in more detail by just the three basic actions—three actions that do, in fact, explain nearly everything else, too.

It is hard to believe that nearly all the vast apparent variety in Nature results from the monotony of repeatedly combining just these three basic actions. But it does. I'll outline a bit of how some of this variety arises.

We may start with photons (see Fig. 70). What is the probability that two photons, at points 1 and 2 in space-time, go to two detectors, at points 3 and 4? There are two main ways this event could happen and each depends on two things happening concomitantly: the photons could go directly—P(1 to 3)*P(2 to 4)—or they could "cross over"—P(1 to 4)*P(2 to 3). The resulting amplitudes for these two possibilities are added, and there is interference (as we saw in the second lecture), making the final arrow vary in length, depending on the relative location of the points in space-time.

What if we make 3 and 4 the same point in space-time (see Fig. 71)? Let's say both photons end up at point 3, and see how this affects the probability of the event. Now we have P(1 to 3)*P(2 to 3) and P(2 to 3)*P(1 to 3), which result in two identical arrows. When added, their sum is twice the length of either one, and produces a final arrow whose square is four times the square of either arrow alone. Because the two arrows are identical, they are always "lined up." In other words, the interference doesn't fluctuate ac-

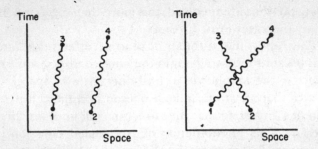

FIGURE 70.   *Photons at points 1 and 2 in space-time have an amplitude to arrive at points 3 and 4 in space-time that is approximated by considering two main ways the event could happen: P(1 to 3) * P(2 to 4) and P(1 to 4) * P(2 to 3), shown above. Depending on the relative locations of points 1, 2, 3, and 4, there are varying degrees of interference.*

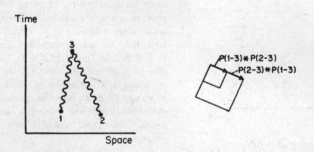

FIGURE 71.   *When points 1 and 3 are made to converge, the two arrows— P(1 to 3) * P(2 to 3) and P(2 to 3) * P(1 to 3)—are identical in length and direction. When they are added they always "line up" and form an arrow with twice the length of either arrow alone, with a square four times as large. Thus photons tend to go to the same point in space-time. This effect is magnified even more by more photons. This is the basis of a laser's operation.*

cording to the relative separation between points 1 and 2; it is always positive. If we didn't think about the always positive interference of the two photons, we should have thought that we would get twice the probability, on average. Instead, we get four times the probability all the time. When many photons are involved, this more-than-expected probability increases even further.

This results in a number of practical effects. We can say that photons tend to get into the same condition, or "state" (the way the amplitude to find one varies in space). The chance that an atom emits a photon is enhanced if some photons (in a state that the atom can emit into) are already present. This phenomenon of "stimulated emission" was discovered by Einstein when he launched the quantum theory proposing the photon model of light. Lasers work on the basis of this phenomenon.

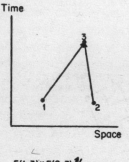

E(1–3)＊E(2–3)//E(2–3)＊E(1–3)

FIGURE 72. *If two electrons (with the same polarization) try to go to the same point in space-time, the interference is always negative because of the effects of polarization: the two identical arrows—E(1 to 3) * E(2 to 3) and E(2 to 3) * E(1 to 3)—are subtracted to make a final arrow of no length. The aversion of two electrons to occupy the same place in space-time is called the "Exclusion Principle," and accounts for the great variety of atoms in the universe.*

If we made the same comparison with our fake, spin-zero electrons, the same thing would happen. But in the real world, where electrons are polarized, something very different happens: the two arrows, E(1 to 3) * E(2 to 4) and E(1 to 4) * E(2 to 3), are subtracted—one of them is turned 180° before they are added. When points 3 and 4 are the same, the two arrows have the same length and direction and thus cancel out when they are subtracted (see Fig. 72). That means electrons, unlike photons, do not like to go to the same place; they avoid each other like the plague—no two electrons with the same polarization can be at the same point in space-time—it's called the "exclusion principle."

This exclusion principle turns out to be the origin of the great variety of chemical properties of the atoms. One proton exchanging photons with one electron dancing around it is called a hydrogen atom. Two protons in the same nucleus exchanging photons with two electrons (polarized in opposite directions) is called a helium atom. You see, the chemists have a complicated way of counting: instead of saying "one, two, three, four, five protons," they say, "hydrogen, helium, lithium, beryllium, boron."

There are only two states of polarization available to electrons, so in an atom with three protons in the nucleus exchanging photons with three electrons—a condition called a lithium atom—the third electron is farther away from the nucleus than the other two (which have used up the nearest available space), and exchanges fewer photons. This causes the electron to easily break away from its own nucleus under the influence of photons from other atoms. A large number of such atoms close together easily lose their individual third electrons to form a sea of electrons swimming around from atom to atom. This sea of electrons reacts to any small electrical force (photons), generating a current of electrons—I am describing lithium metal conducting electricity. Hydrogen and helium atoms do not lose their electrons to other atoms. They are "insulators."

All the atoms—more than one hundred different kinds—are made up of a certain number of protons exchanging photons with the same number of electrons. The patterns in which they gather are complicated and offer an enormous variety of properties: some are metals, some are insulators, some are gases, others are crystals; there are soft things, hard things, colored things, and transparent things—a terrific cornucopia of variety and excitement that comes from the exclusion principle and the repetition again and again and again of the three very simple actions P(A to B), E(A to B), and j. (If the electrons in the world were unpolarized, all the atoms would have very similar properties: the electrons would all cluster together, close to the nucleus of their own atom, and would not be easily attracted to other atoms to make chemical reactions.)

You might wonder how such simple actions could produce such a complex world. It's because phenomena we see in the world are the result of an enormous intertwining of tremendous numbers of photon exchanges and interferences. Knowing the three fundamental actions is only a very small beginning toward analyzing any *real* situation, where there is such a multitude of photon exchanges going on that it is impossible to calculate—experience has to be gained as to which possibilities are more important. Thus we invent such ideas as "index of refraction" or "compressibility" or "valence" to help us calculate in an approximate way when there's an enormous amount of detail going on underneath. It's analogous to knowing the rules of chess—which are fundamental and simple—compared to being able to play chess well, which involves understanding the character of each position and the nature of various situations—which is much more advanced and difficult.

The branches of physics that deal with questions such as why iron (with 26 protons) is magnetic, while copper (with 29) is not, or why one gas is transparent and another one is not, are called "solid-state physics," or "liquid-state physics," or "honest physics." The branch of physics that found these three simple little actions (the easiest part) is called "fundamental physics"—we stole that name in order to make the other physicists feel uncomfortable! The most interesting problems today—and certainly the most practical problems—are obviously in solid-state physics. But someone said there is nothing so practical as a good theory, and the theory of quantum electrodynamics is definitely a good theory!

Finally, I would like to return to that number 1.00115965221, the number that I told you about in the first lecture that has been measured and calculated so carefully. The number represents the response of an electron to an external magnetic field—something called the "magnetic moment." When Dirac first worked out the rules to calculate this number, he used the formula for E(A to B) and got a very simple answer, which we will consider in our units as 1. The diagram for this first approximation of the magnetic moment of an electron is very simple—an electron goes from place to place in space-time and couples with a photon from a magnet (see Fig. 73).

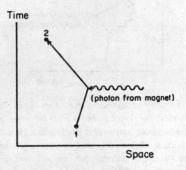

FIGURE 73.   *The diagram for Dirac's calculation of the magnetic moment of an electron is very simple. The value represented by this diagram will be called 1.*

After some years it was discovered that this value was not exactly 1, but slightly more—something like 1.00116. This correction was worked out for the first time in 1948 by Schwinger as $j*j$ divided by 2 pi, and was due to an alternative way the electron can go from place to place: instead of going directly from one point to another, the electron goes along for a while and suddenly emits a photon; then (horrors!) it absorbs its own photon (see Fig. 74). Perhaps there's something "immoral" about that, but the electron does it! To calculate the arrow for this alternative, we have to make an arrow for every place in space-time that the photon can be emitted and every place it can be absorbed. Thus there will be two extra E(A to B)'s, a P(A to B) and two extra j's, all multiplied together. Students learn how to do this simple calculation in their elementary quantum electrodynamics course, in their second year of graduate school.

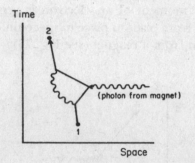

FIGURE 74. *Laboratory experiments show that the actual value of the magnetic moment of an electron is not 1, but a little bit more. This is because there are alternatives: the electron can emit a photon and then absorb it—requiring two extra E(A to B)'s, a P(A to B), and two extra j's. Schwinger calculated the adjustment that takes this alternative into account to be j*j divided by 2 pi. Since this alternative is indistinguishable experimentally from the original way the electron can go—an electron starts at point 1 and ends up at point 2—the arrows for the two alternatives are added, and there is interference.*

But wait: experiments have measured the behavior of an electron so accurately that we have to consider still other possibilities in our calculations—all the ways the electron can go from place to place with *four* extra couplings (see Fig. 75). There are three ways the electron can emit and absorb two photons. There's also a new, interesting possibility (shown at the right of Fig. 75): one photon is emitted; it makes a positron-electron pair, and—again, if you'll hold

your "moral" objections—the electron and positron annihilate, creating a new photon that is ultimately absorbed
by the electron. That possibility also has to be figured in!

It took two "independent" groups of physicists two years
to calculate this next term, and then another year to find

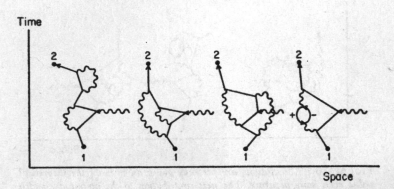

FIGURE 75. *Laboratory experiments became so accurate that further alternatives, involving four extra couplings (over all possible intermediate points
in space-time), had to be calculated, some of which are shown here. The
alternative on the right involves a photon disintegrating into a positron-
electron pair (as described in Fig. 64), which annihilates to form a new photon,
which is ultimately absorbed by the electron.*

out there was a mistake—experimenters had measured the
value to be slightly different, and it looked for awhile that
the theory didn't agree with experiment for the first time,
but no: it was a mistake in arithmetic. How could two
groups make the same mistake? It turns out that near the
end of the calculation the two groups compared notes and
ironed out the differences between their calculations, so
they were not really independent.

The term with *six* extra *j*'s involves even more possible
ways the event can happen, and I'll draw a few of them for
you now (see Fig. 76). It took twenty years to get this extra
accuracy figured into the theoretical value of the magnetic

moment of an electron. Meanwhile the experimenters made even more detailed experiments and added a few more digits onto their number—and the theory still agreed with it.

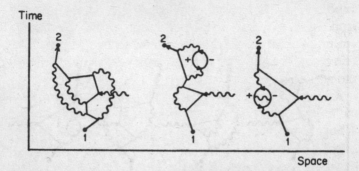

FIGURE 76. *Calculations are presently going on to make the theoretical value even more accurate. The next contribution to the amplitude, which represents all possibilities with six extra couplings, involves something like 70 diagrams, three of which are shown here. As of 1983, the theoretical number was 1.00115965246, with an uncertainty of about 20 in the last two digits; the experimental number was 1.00115965221, with an uncertainty of about 4 in the last digit. This accuracy is equivalent to measuring the distance from Los Angeles to New York, a distance of over 3,000 miles, to within the width of a human hair.*

So, to make our calculations we make these diagrams, write down what they correspond to mathematically, and add the amplitudes—a straightforward, "cookbook" process. Therefore, it can be done by machines. Now that we have super-duper computers, we have begun to compute the term with eight extra j's. At the present time the theoretical number is 1.00115965246; experimentally, it's 1.00115965221, plus or minus 4 in the last decimal place. Some of the uncertainty in the theoretical value (about 4 in the last decimal place) is due to the computer's rounding off numbers; most of it (about 20) is due to the fact that the value for j is not exactly known. The term for eight extra j's involves something like nine hundred diagrams, with a hundred thousand terms each—a fantastic calculation—and it's being done right now.

I am sure that in a few more years, the theoretical and experimental numbers for the magnetic moment of an electron will be worked out to still more places. Of course, I am not sure whether the two values will still agree. That, one can never tell until one makes the calculation and does the experiments.

And so we have come full circle to the number I chose to "intimidate" you with at the beginning of these lectures. I hope you understand the significance of this number much better now: it represents the extraordinary degree to which we've been constantly checking that the strange theory of quantum electrodynamics is indeed correct.

Throughout these lectures I have delighted in showing you that the price of gaining such an accurate theory has been the erosion of our common sense. We must accept some very bizarre behavior: the amplification and suppression of probabilities, light reflecting from all parts of a mirror, light travelling in paths other than a straight line, photons going faster or slower than the conventional speed of light, electrons going backwards in time, photons suddenly disintegrating into a positron-electron pair, and so on. That we must do, in order to appreciate what Nature is really doing underneath nearly all the phenomena we see in the world.

With the exception of technical details of polarization, I have described to you the framework by which we understand all these phenomena. We draw *amplitudes* for every way an event can happen and add them when we would have expected to add probabilities under ordinary circumstances; we multiply amplitudes when we would have expected to multiply probabilities. Thinking of everything in terms of amplitudes may cause difficulties at first because of their abstraction, but after a while, one gets used to this strange language. Underneath so many of the phenomena we see every day are only three basic actions: one is described by the simple coupling number, j; the other two by functions—P(A to B) and E(A to B)—both of which are closely related. That's all there is to it, and from it all the rest of the laws of physics come.

However, before I finish this lecture, I would like to make a few additional remarks. One can understand the spirit and character of quantum electrodynamics without including this technical detail of polarization. But I'm sure you'll all feel uncomfortable unless I say something about what I've been leaving out. Photons, it turns out, come in four different varieties, called polarizations, that are related geometrically to the directions of space and time. Thus there are photons polarized in the X, Y, Z, and T directions. (Perhaps you have heard somewhere that light comes in only two states of polarization—for example, a photon going in the Z direction can be polarized at right angles, either in the X or Y direction. Well, you guessed it: in situations where the photon goes a long distance and appears to go at the speed of light, the amplitudes for the Z and T terms exactly cancel out. But for virtual photons going between a proton and an electron in an atom, it is the T component that is the most important.)

In a similar manner, an electron can be in one of four conditions that are also related to geometry, but in a somewhat more subtle manner. We can call these conditions 1, 2, 3, and 4. Calculating the amplitude for an electron going from point A to point B in space-time becomes somewhat more complicated, because we can now ask questions such as, "What is the amplitude that an electron liberated in condition 2 at the point A arrives in condition 3 at the point B?" The sixteen possible combinations—coming from the four different conditions an electron can start in at A and the four different conditions it can end up in at B—are related in a simple mathematical way to the formula for that E(A to B) I told you about.

For a photon, no such modification is necessary. Thus a photon polarized in the X direction at A will still be polarized in the X direction at B, arriving with the amplitude P(A to B).

Polarization produces a large number of different possible couplings. We could ask, for example, "What is the amplitude that an electron in condition 2 absorbs a photon polarized in the X direction and thereby turns into an electron in condition 3?" All the possible combinations of polarized electrons and photons do not couple, but those that do, do so with the same amplitude j, but sometimes with an additional turn of the arrow by some multiple of 90°.

These possibilities for the different kinds of polarization and the nature of the couplings can all be deduced in a very elegant and beautiful manner from the principles of quantum electrodynamics and two further assumptions: 1) the results of an experiment are not affected if the apparatus with which you are making experiments is turned in some other direction, and 2) it also doesn't make any difference if the apparatus is in a spaceship moving at some arbitrary speed. (This is the principle of relativity.)

This elegant and general analysis shows that every particle must be in one or another class of possible polarizations, which we call spin 0, spin 1/2, spin 1, spin 3/2, spin 2, and so on. The different classes behave in different ways. A spin 0 particle is the simplest—it has just one component, and is not effectively polarized at all. (The fake electrons and photons that we have been considering in this lecture are spin 0 particles. So far, no fundamental spin 0 particles have been found.) A real electron is an example of a spin 1/2 particle, and a real photon is an example of a spin 1 particle. Both spin 1/2 and spin 1 particles have four components. The other types would have more components, such as spin 2 particles, with ten components.

I said that the connection between relativity and polarization is simple and elegant, but I'm not sure I can explain it simply and elegantly! (It would take me at least one additional lecture to do it.) Although the details of polarization are not essential to understanding the spirit and character of quantum electrodynamics, they are, of course, essential to the correct calculation of any real process, and often have profound effects.

In these lectures we have been concentrating on relatively simple interactions between electrons and photons at very small distances, in which only a few particles are involved. But I would like to make one or two remarks about how these interactions appear in the larger world, where very, very large numbers of photons are being exchanged. On such a large scale, the calculation of arrows gets very complicated.

There are, however, some situations that are not so difficult to analyze. There are circumstances, for example, where the amplitude to emit a photon by a source is independent of whether another photon has been emitted. This can happen when the source is very heavy (the nucleus of an atom), or when a very large number of electrons are all moving the same way, such as up and down in the antenna of a broadcasting station or going around in the coils of an electromagnet. Under such circumstances a large number of photons are emitted, all of exactly the same kind. The amplitude of an electron to absorb a photon in such an environment is independent of whether it or any other electron has absorbed other photons before. Therefore its entire behavior can be given by just this amplitude for an electron to absorb a photon, which depends only on the electron's position in space and time. Physicists use ordinary words to describe this circumstance. They say the electron is moving in an external field.

Physicists use the word "field" to describe a quantity that depends on position in space and time. Temperatures in the air provide a good example: they vary according to where and when you make your measurements. When we take polarization into account, there are more components to the field. (There are four components—corresponding to the amplitude to absorb each of the different kinds of polarization (X, Y, Z, T) the photon might be in—technically called the vector and scalar electromagnetic potentials. From combinations of these, classical physics derives more convenient components called the electric and magnetic fields.)

In a situation where the electric and magnetic fields are varying slowly enough, the amplitude for an electron to travel over a very long distance depends on the path it takes. As we saw earlier in the case of light, the most important paths are the ones where the angles of the amplitudes from nearby paths are nearly the same. The result is that the particle doesn't necessarily go in a straight line.

This brings us all the way back to classical physics, which supposes that there are fields and that electrons move through them in such a way as to make a certain quantity least. (Physicists call this quantity "action" and formulate this rule as the "principle of least action.") This is one example of how the rules of quantum electrodynamics produce phenomena on a large scale. We could expand in many directions from here, but we have to limit the scope of these lectures somewhere. I just wanted to remind you that the effects that we see on a large scale and the strange phenomena we see on a small scale are both produced by the interaction of electrons and photons, and are all described, ultimately, by the theory of quantum electrodynamics.

4

Loose Ends

I am going to divide this lecture into two parts. First, I am going to talk about problems associated with the theory of quantum electrodynamics itself, supposing that all there is in the world is electrons and photons. Then I will talk about the relation of quantum electrodynamics to the rest of physics.

The most shocking characteristic of the theory of quantum electrodynamics is the crazy framework of amplitudes, which you might think indicates problems of some sort! However, physicists have been fiddling around with amplitudes for more than fifty years now, and have gotten very used to it. Furthermore, all the new particles and new phenomena that we are able to observe fit perfectly with everything that can be deduced from such a framework of amplitudes, in which the probability of an event is the square of a final arrow whose length is determined by combining arrows in funny ways (with interferences, and so on). So this framework of amplitudes has *no experimental doubt* about it: you can have all the philosophical worries you want as to what the amplitudes mean (if, indeed, they mean anything at all), but because physics is an experimental science and the framework agrees with experiment, it's good enough for us so far.

There is a set of problems associated with the theory of quantum electrodynamics that has to do with improving the method of calculating the sum of all the little arrows—

various techniques that are available in different circumstances—that take the graduate students three or four years to master. Since they are technical problems, I am not going to discuss them here. It's just a matter of continuously improving the techniques for analyzing what the theory really has to say in different circumstances.

But there is one additional problem that is characteristic of the theory of quantum electrodynamics itself, which took twenty years to overcome. It has to do with ideal electrons and photons and the numbers n and j.

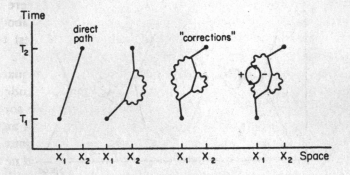

FIGURE 77. *When we calculate the amplitude for an electron to go from point to point in space-time, we use the formula for E(A to B) for the direct path. (Then we make "corrections" that include one or more photons being emitted and absorbed.) E(A to B) depends on $(X_2 - X_1)$, $(T_2 - T_1)$ and n, a number we stick into the formula to make the answer come out right. The number n is called the "rest-mass" of an "ideal" electron, and cannot be measured experimentally because the rest-mass of a real electron, m, includes all the "corrections." There is a certain difficulty in calculating the n to be used in E(A to B), that took twenty years to overcome.*

If electrons were ideal, and went from point to point in space-time *only* by the direct path (shown at the left in Fig. 77), then there would be no problem: n would simply be the mass of an electron (which we can determine by observation), and j would simply be its "charge" (the amplitude for the electron`to couple with a photon). It can also be determined by experiment.

But no such ideal electrons exist. The mass we observe in the laboratory is that of a *real* electron, which emits and absorbs its own photons from time to time, and therefore depends on the amplitude for coupling, j. And the charge we observe is between a *real* electron and a *real* photon—which can form an electron-positron pair from time to time—and therefore depends on E (A to B), which involves n (see Fig. 78). Since the mass and charge of an electron

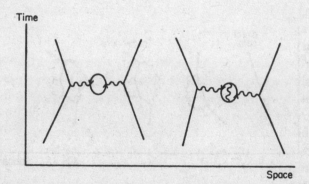

FIGURE 78. *The experimentally measured amplitude for an electron to couple with a photon, a mysterious number, e, is a number determined by experiment that includes all the "corrections" for a photon going from point to point in space-time, of which two are shown here. When calculating, we need a number, j, that does not include these corrections, but includes only the photon going directly from point to point. A difficulty exists with computing this j that is similar to the difficulty in computing the value of n.*

are affected by these and all other alternatives, the experimentally measured mass, m, and the experimentally measured charge, e, of the electron are different from the numbers we use in our calculations, n and j.

If there were a definite mathematical connection between n and j on the one hand, and m and e on the other, there would still be no problem: we would simply calculate what values of n and j we need to start with in order to end up with the observed values, m and e. (If our calculations didn't agree with m and e, we would jiggle the original n and j around until they did.)

Let's see how we actually calculate m. We write a series of terms that is something like the series we saw for the magnetic moment of the electron: the first term has no couplings—just E (A to B)—and represents an ideal electron going directly from point to point in space-time. The second term has two couplings and represents a photon being emitted and absorbed. Then come terms with four, six, and eight couplings, and so on (some of these "corrections" are shown in Fig. 77).

When calculating terms with couplings, we must consider (as always) all the possible points where couplings can occur, right down to cases where the two coupling points are on top of each other—with zero distance between them. The problem is, when we try to calculate all the way down to zero distance, the equation blows up in our face and gives meaningless answers—things like infinity. This caused a lot of trouble when the theory of quantum electrodynamics first came out. People were getting infinity for every problem they tried to calculate! (One should be able to go down to zero distance in order to be mathematically consistent, but that's where there is no n or j that makes any sense; that's where the trouble is.)

Well, instead of including all possible coupling points down to a distance of zero, if one *stops* the calculation when the distance between coupling points is very small—say, 10^{-30} centimeters, billions and billions of times smaller than anything observable in experiment (presently 10^{-16} centimeters)—then there are definite values for n and j that we can use so that the calculated mass comes out to match the m observed in experiments, and the calculated charge matches the observed charge, e. Now, here's the catch: if somebody else comes along and stops their calculation at a different distance—say, 10^{-40} centimeters—*their* values for n and j needed to get the same m and e come out *different*!

Twenty years later, in 1949, Hans Bethe and Victor Weisskopf noticed something: if two people who stopped at different distances to determine n and j from the same m and e then calculated the answer to some *other* problem— each using the appropriate but different values for n and j—when all the arrows from all the terms were included, their answers to this other problem came out nearly the same! In fact, the closer to zero distance that the calculations for n and j were stopped, the better the final answers for the other problem would agree! Schwinger, Tomonaga, and I independently invented ways to make definite calculations to confirm that it is true (we got prizes for that). People could finally calculate with the theory of quantum electrodynamics!

So it appears that the *only* things that depend on the small distances between coupling points are the values for n and j—*theoretical numbers that are not directly observable anyway*; everything else, which *can* be observed, seems not to be affected.

The shell game that we play to find n and j is technically called "renormalization." But no matter how clever the word, it is what I would call a dippy process! Having to resort to such hocus-pocus has prevented us from proving that the theory of quantum electrodynamics is mathematically self-consistent. It's surprising that the theory still hasn't been proved self-consistent one way or the other by now; I suspect that renormalization is not mathematically legitimate. What *is* certain is that we do not have a good mathematical way to describe the theory of quantum electrodynamics: such a bunch of words to describe the connection between n and j and m and e is not good mathematics.[1]

There is a most profound and beautiful question associated with the observed coupling constant, e—the amplitude for a real electron to emit or absorb a real photon. It is a simple number that has been experimentally determined to be close to -0.08542455. (My physicist friends won't recognize this number, because they like to remember it as the inverse of its square: about 137.03597 with an uncertainty of about 2 in the last decimal place. It has been a mystery ever since it was discovered more than fifty years ago, and all good theoretical physicists put this number up on their wall and worry about it.)

Immediately you would like to know where this number for a coupling comes from: is it related to pi, or perhaps to the base of natural logarithms? Nobody knows. It's one of the *greatest* damn mysteries of physics: a *magic number* that comes to us with no understanding by man. You might say the "hand of God" wrote that number, and "we don't know how He pushed His pencil." We know what kind of a dance to do experimentally to measure this number very accurately, but we don't know what kind of a dance to do on a computer to make this number come out—without putting it in secretly!

A good theory would say that e is the square root of 3

---

[1] Another way of describing this difficulty is to say that perhaps the idea that two points can be infinitely close together is wrong—the assumption that we can use geometry down to the last notch is false. If we make the minimum possible distance between two points as small as 10^{-100} centimeters (the smallest distance involved in any experiment today is around 10^{-16} centimeters), the infinities disappear, all right—but other inconsistencies arise, such as the total probability of an event adds up to slightly more or less than 100%, or we get negative energies in infinitesimal amounts. It has been suggested that these inconsistencies arise because we haven't taken into account the effects of gravity—which are normally very, very weak, but become important at distances of 10^{-33} cm.

over 2 pi squared, or something. There have been, from time to time, suggestions as to what *e* is, but none of them has been useful. First, Arthur Eddington proved by pure logic that the number the physicists like had to be exactly 136, the experimental number at that time. Then, as more accurate experiments showed the number to be closer to 137, Eddington discovered a slight error in his earlier argument, and showed by pure logic again that the number had to be the integer 137! Every once in a while, someone notices that a certain combination of pi's and e's (the base of the natural logarithms), and 2's and 5's produces the mysterious coupling constant, but it is a fact not fully appreciated by people who play with arithmetic that you would be surprised how *many* numbers you can make out of pi's and e's and so on. Therefore, throughout the history of modern physics, there has been paper after paper by people who have produced an *e* to several decimal places, only to have the next round of improved experiments disagree with it.

Even though we have to resort to a dippy process to calculate *j* today, it's possible that someday a legitimate mathematical connection between *j* and *e* will be found. That would mean that *j* is the mysterious number, and from it comes *e*. In such a case there would doubtless be another batch of papers that tell us how to calculate *j* "with our bare hands," so to speak, proposing that *j* is 1 divided by 4 * pi, or something.

That exposes all the problems associated with quantum electrodynamics.

When I planned these lectures, I intended to concentrate only on the part of physics that we know very well—to describe it fully and to say no more. But now that we've come this far, being a professor (which means having the habit of not being able to stop talking at the right time), I cannot resist telling you something about the rest of physics.

First, I must immediately say that the rest of physics has not been checked anywhere nearly as well as electrodynamics: some of the things I'm going to tell you are good guesses, some are partly worked-out theories, and others are pure speculation. Therefore this presentation is going to look like a relative mess, compared to the other lectures; it will be incomplete and lacking in many details. Nevertheless, it turns out that the structure of the theory of QED serves as an excellent basis for describing other phenomena in the rest of physics.

I'll begin by talking about protons and neutrons, which make up the nuclei of atoms. When protons and neutrons were first discovered it was thought that they were simple particles, but very soon it became clear that they were not simple—simple in the sense that their amplitude to go from one point to another could be explained by the formula E (A to B), but with a different number for n stuck in. For example, the proton has a magnetic moment that, if calculated in the same way as for the electron, should be close to 1. But in fact, experimentally it comes out completely crazy—2.79! Therefore it was soon realized that something's going on inside the proton that is not accounted for in the equations of quantum electrodynamics. And the neutron, which should have no magnetic interaction at all if it is really neutral, has a magnetic moment of about −1.93! So it was known for a long time that something fishy is going on inside the neutron as well.

There was also the problem of what holds the neutrons and protons together inside the nucleus. It was realized right away that it could not be the exchange of photons, because the forces holding the nucleus together were much stronger—the energy required to break up a nucleus is much greater than that required to knock an electron away from an atom in the same proportion that an atomic bomb is more destructive than dynamite: exploding dynamite is a rearrangement of the electron patterns, while an exploding atomic bomb is a rearrangement of the proton-neutron patterns.

To find out more about what holds the nuclei together, many experiments were made in which protons with higher and higher energies were smashed into nuclei. It was expected that only protons and neutrons would come out. But when the energies became sufficiently large, new particles came out. First there were pions, then lambdas, and sigmas, and rhos, and they ran out of the alphabet. Then came particles with numbers (their masses), such as sigma 1190 and sigma 1386. It soon became clear that the number of particles in the world was open-ended, and depended on the amount of energy used to break apart the nucleus. There are over four hundred such particles at present. We can't accept four hundred particles; that's too complicated![2]

Great inventors like Murray Gell-Mann nearly went crazy trying to figure out the rules by which all these particles behave, and in the early 1970s they came up with the quantum theory of strong interactions (or "quantum chromodynamics"), whose main actors are particles called "quarks." All of the particles made of quarks come in two classes: some, like the proton and neutron, are made out of three quarks (and go by the horrible name of "baryons"); others, such as the pions, are made of a quark and an anti-quark (and are called "mesons").

Let me make a table of the fundamental particles as they appear today (see Fig. 79). I'll begin with the particles that go from point to point according to the formula E(A to B)—modified by the same kind of polarization rules as an

_____

[2] Although many particles come out of the nucleus in high-energy experiments, in low-energy experiments—in more normal conditions—the nuclei are found to contain only protons and neutrons.

electron—called "spin ·1/2" particles. The first of these particles is the electron, and its mass number is 0.511 in units that we use all the time, called MeV.[3]

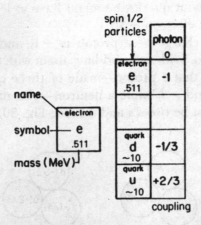

FIGURE 79. *Our list of all the particles in the world begins with "spin 1/2" particles: the electron (with a mass of 0.511 MeV), and two "flavors" of quarks, d and u (both with a mass of about 10 MeV). Electrons and quarks have a "charge"—that is, they couple with photons in the following amounts (in terms of the coupling constant,* $-j$*):* -1*,* $-1/3$*, and* $+2/3$*.*

Under the electron I will leave a space (to be occupied later), and under that I will list two types of quarks—the *d* and the *u*. The mass of these quarks is not exactly known; a good guess is around 10 MeV for each one. (The neutron is slightly heavier than the proton, which seems to imply—as you will see in a moment—that the *d* quark is somewhat heavier than the *u* quark.)

Next to each particle I will list its charge, or coupling constant, in terms of $-j$, the number for couplings with photons with its sign reversed. This makes the charge for

[3] An MeV is very small—appropriate to such particles—about 1.78 • 10^{-27} grams.

the electron −1, consistent with a convention started by Benjamin Franklin that we've been stuck with ever since. For the *d* quark the amplitude to couple with a photon is −1/3, and for the *u* quark it is +2/3. (Had Benjamin Franklin known about quarks, he might have at least made the charge of an electron −3!)

Now, the charge of a proton is +1, and a neutron's charge is zero. With some fiddling about with the numbers, you can see that a proton—made of three quarks—must be two *u*'s and a *d*, while a neutron—also made of three quarks—must be two *d*'s and a *u* (see Fig. 80).

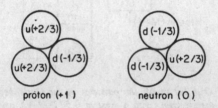

FIGURE 80. *All particles made of quarks come in one of only two possible classes: those made of a quark and an anti-quark, and those made of three quarks, of which the proton and the neutron are the most common examples. The charge of the* d *and* u *quarks combine to make* +1 *for the proton and zero for the neutron. The fact that the proton and neutron are made of charged particles going around inside them gives a clue as to why the proton has a magnetic moment higher than 1, and why the supposedly neutral neutron has a magnetic moment at all.*

What holds the quarks together? Is it the photons going back and forth? (Because a *d* quark has a charge of −1/3 and a *u* quark has a charge of +2/3, quarks, as well as electrons, emit and absorb photons.) No, these electrical forces are far too weak to do that. Something else has been invented to go back and forth and hold quarks together; something called "gluons."[4] Gluons are an example of an-

⁴ Notice the names: "photon" comes from the Greek word for light;

other type of particle called "spin 1" (as are photons); they go from point to point with an amplitude determined by exactly the same formula as for photons, P(A to B). The amplitude for gluons to be emitted or absorbed by quarks is a mysterious number, g, that is much larger than j (see Fig. 81).

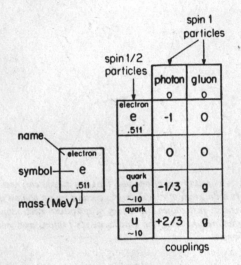

FIGURE 81. *"Gluons" hold quarks together to make protons and neutrons, and indirectly account for the fact that protons and neutrons hold themselves together in the nucleus of an atom. Gluons hold quarks together with forces much stronger than electrical forces. The coupling constant of gluons, g, is much larger than j, which makes the calculation of terms with couplings in them much more difficult: the best accuracy that can be hoped for so far is only 10%.*

"electron" comes from the Greek word for amber, the beginning of electricity. But as modern physics has progressed, the names of the particles have shown a deteriorating interest in classical Greek until we make up such words as "gluons." Can you guess why they're called "gluons?" in fact, d and u stand for words, but I don't want to confuse you—a d quark is no more "down" than a u quark is "up." Incidentally, the d-ness or u-ness of a quark is called its "flavor."

The diagrams we make of quarks exchanging gluons are very similar to the pictures we draw for electrons exchanging photons (see Fig. 82). So similar, in fact, that you might say that the physicists have no imagination—that they just copied the theory of quantum electrodynamics for the strong interactions! And you're right: that's what we did, but with a little twist.

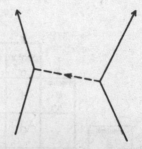

FIGURE 82. *The diagram of one way that two quarks can exchange a gluon is so similar to a diagram of two electrons exchanging a photon that you might think the physicists just copied the theory of quantum electrodynamics for the "strong interactions" holding the quarks inside protons and neutrons. Well, they did—almost.*

The quarks have an additional type of polarization that is not related to geometry. The idiot physicists, unable to come up with any wonderful Greek words anymore, call this type of polarization by the unfortunate name of "color," which has nothing to do with color in the normal sense. At a particular time, a quark can be in one of three conditions, or "colors"—R, G, or B (can you guess what they stand for?). A quark's "color" can be changed when the quark emits or absorbs a gluon. The gluons come in eight different types, according to the "colors" they can couple with. For example, if a red quark changes to green, it emits a red-antigreen gluon—a gluon that takes the red from the quark and gives it green ("antigreen" means the gluon is carrying green in the opposite direction). This gluon could be absorbed by a green quark, which changes to red (see Fig. 83). There are eight different possible gluons, such as red-antired, red-antiblue, red-antigreen, and so on (you'd think there'd be nine, but for technical reasons, one is missing). The theory is not very complicated. The complete rule of gluons is: gluons couple with things having "color"— it just requires a little bookkeeping to keep track of where the "colors" go.

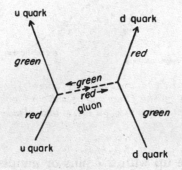

FIGURE 83. *Gluon theory differs from electrodynamics in that gluons couple with things that are "colored" (in one of three possible conditions—"red," "green," and "blue"). Here, a red u quark changes to green by emitting a red-antigreen gluon that is absorbed by a green d quark changing to red. (If the "color" is being carried backwards in time, it takes the prefix "anti.")*

There is, however, an interesting possibility created by this rule: gluons can couple with other gluons (see Fig. 84). For instance, a green-antiblue gluon meeting a red-anti-green gluon results in a red-antiblue gluon. Gluon theory is very simple—you just make the diagram and follow the "colors." The strengths of the couplings in all the diagrams is determined from the coupling constant for gluons, g.

Gluon theory is really not a great deal different in form from quantum electrodynamics. How, then, does it compare with experiment? For example, how does the observed magnetic moment of the proton compare with the value calculated from the theory?

The experiments are very accurate—they show the magnetic moment to be 2.79275. At the very best, the theory

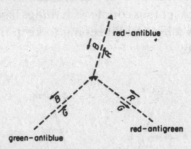

FIGURE 84.   *Since gluons are themselves "colored," they can couple to each other. Here a green-antiblue gluon couples with a red-antigreen gluon to form a red-antiblue gluon. Gluon theory is easy to understand—you just follow the "colors."*

can only come up with 2.7 plus or minus 0.3—if you're sufficiently optimistic about the accuracy of your analysis— an error of 10% which is 10,000 times less accurate than experiment! We have a simple, definite theory that is supposed to explain all the properties of protons and neutrons, yet we can't calculate anything with it, because the mathematics is too hard for us. (You can guess what I'm working on, and I'm not getting anywhere.) The reason we can't calculate to any great accuracy is because the coupling constant for gluons, g, is so much larger than for electrons. Terms with two, four, and even six couplings are not just minor corrections to the main amplitude; they represent considerable contributions that can't be ignored. Thus there are arrows from so many different possibilities that we haven't been able to organize them in a reasonable way to find out what the final arrow is.

In books it says that science is simple: you make up a theory and compare it to experiment; if the theory doesn't work, you throw it away and make a new theory. Here we have a definite theory and hundreds of experiments, but we can't compare them! It's a situation that has never before existed in the history of physics. We're boxed in, temporarily, unable to come up with a method of calculation. We're snowed under by all the little arrows.

Despite our difficulties in calculating with the theory, we do understand some things qualitatively about quantum chromodynamics (strong interactions of quarks and gluons). The objects made of quarks that we see are "colored" neutral: groups of three quarks contain one quark of each "color," and quark-antiquark pairs have an equal amplitude to be red-antired, green-antigreen, or blue-antiblue. We also understand why quarks can never be produced as individual particles—why, no matter how much energy is used to hit a nucleus against a proton, instead of seeing individual quarks come out, we see a jet of mesons and baryons (quark-antiquark pairs and groups of three quarks).

Quantum chromodynamics and quantum electrodynamics aren't all there is to physics. According to them, a quark cannot change its "flavor": once a u quark, always a u quark; once a d quark, always a d quark. But Nature behaves differently, sometimes. There is a form of radioactivity that happens slowly—the kind that people worry about leaking out of nuclear reactors—called beta decay, which involves a neutron changing into a proton. Since a neutron consists of two d's and a u-type quark while a proton is made of two u's and a d, what really happens is that one of the neutron's d-type quarks changes into a u-type quark (see Fig. 85). Here's how it happens: the d quark emits a new thing like a photon called a W, which has a coupling with an electron and with another new particle called an anti-neutrino, a neutrino going backwards in time. The neutrino is another spin 1/2 type particle (like the electron and the quarks), but it has no mass and no charge (it does not interact with photons). It also does not interact with gluons; it only couples with the W (see Fig. 86).

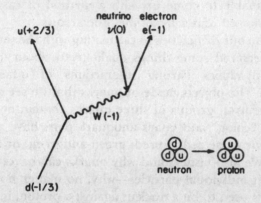

FIGURE 85. *When a neutron disintegrates into a proton (a process called "beta decay"), the only thing that changes is the "flavor" of one quark—from d to u—with an electron and an anti-neutrino coming out. This process happens relatively slowly, so an intermediate particle (called a "W-interme-diate-boson") with a very high mass (about 80,000 MeV) and a charge of −1 was proposed.*

The W is a spin 1 type particle (like the photon and the gluon), that changes the "flavor" of a quark and takes away its charge—the d, charged $-1/3$, changes into a u, charged $+2/3$, a difference of -1. (It doesn't change the quark's "color.") Because the W_- takes away a charge of -1 (and its anti-particle, the W_+, takes away a charge of $+1$), it can also couple with a photon. Beta decay takes much longer than the interactions of photons and electrons, so it is thought that the W must have a very high mass (about 80,000 MeV), unlike the photon and gluon. We have not been able to see

the W by itself because of the very high energy required
to knock loose a particle with such a very high mass.[5]

There is another particle, which we could think of as a
neutral W, called Z_0. The Z_0 does not change the charge
of a quark, but does couple with a d quark, a u quark, an

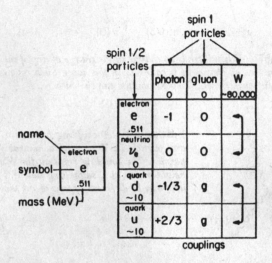

FIGURE 86. *The W couples with the electron and neutrino on the one hand,
and the* d *and* u *quark on the other.*

electron, or a neutrino (see Fig. 87). This interaction has
the misleading name of "neutral currents," and caused a
lot of excitement when it was discovered a few years ago.

The theory of W's is nice and neat if you allow for a
three-way coupling between the three types of W's (see Fig.
88). The observed coupling constant for W's is much the
same as that for the photon—in the neighborhood of j.

---

[5] After these lectures were given, high enough energies were achieved
to produce a W by itself, and its mass was measured to be very close to
the value predicted by the theory.

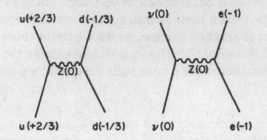

FIGURE 87. *When there is no change in the charge of any of the particles, the W also has no charge (it is called Z_0 in this case). Such interactions are called "neutral currents." Two possibilities are shown here.*

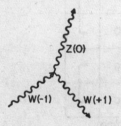

FIGURE 88. *A coupling between a W_{-1}, its anti-particle (a W_{+1}, and a neutral W (Z_0) is possible. The coupling constant for W's is in the neighborhood of j, suggesting that W's and photons may be different aspects of the same thing.*

Therefore the possibility exists that the three W's and the photon are all different aspects of the same thing. Stephen Weinberg and Abdus Salam tried to combine quantum electrodynamics with what's called the "weak interactions" (interactions with W's) into one quantum theory, and they did it. But if you just look at the results they get you can see the glue, so to speak. It's very clear that the photon and the three W's are interconnected somehow, but at the present level of understanding, the connection is difficult to see clearly—you can still see the "seams" in the theories; they have not yet been smoothed out so that the connection becomes more beautiful and, therefore, probably more correct.

So there you are: quantum theory has three main types

of interaction—the "strong interactions" of quarks and gluons, the "weak interactions" of the W's, and the "electrical interactions" of photons. The only particles in the world (according to this picture) are quarks (in "flavors" u and d with three "colors" each), gluons (eight combinations of R, G, and B), W's (charged ± 1 and 0), neutrinos, electrons, and photons—about twenty different particles of six different types (plus their anti-particles). That's not so bad—about twenty different particles—except that's not all.

As nuclei were hit with protons of higher and higher energies, new particles kept coming out. One such particle was the muon, which is in every way exactly the same as the electron, except that its mass is much higher—105.8 MeV, compared to 0.511 for the electron, or about 206 times heavier. It's just as if God wanted to try out a different number for the mass! All of the properties of the muon are completely describable by the theory of electrodynamics—the coupling constant j is the same and E(A to B) is the same; you just put in a different value for n.[6]

Because the muon has a mass about 200 times higher than the electron, the "stopwatch hand" for a muon turns 200 times more rapidly than that of an electron. This has enabled us to test whether electrodynamics still behaves according to the theory at distances 200 times smaller than we've been able to test before—although these distances

---

[6] The magnetic moment of a muon has been measured very accurately—it has been found to be 1.001165924 (with an uncertainty of 9 in the last digit), while the value for the electron is 1.00115965221 (with an uncertainty of 3 in the last digit). You might be curious as to why the magnetic moment of the muon is slightly higher than that of the electron. One of the diagrams we drew had the electron emitting a photon that disintegrates into a positron-electron pair (see Fig. 89). There is also a small amplitude that the emitted photon could make a muon-antimuon pair, which is heavier than the original electron. This is unsymmetrical, because when the muon emits a photon, if that photon makes a positron-electron pair, that pair is *lighter* than the original muon. The theory of quantum electrodynamics accurately describes *every* electrical property of the muon as well as the electron.

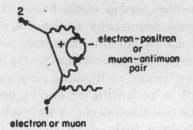

FIGURE 89. *In the process of bombarding nuclei with protons of higher and higher energy, new particles appear. One of these particles is the muon, or heavy electron. The theory describing the muon's interactions is exactly the same as for the electron, except that you just put in a higher number for n into E(A to B). The magnetic moment of a muon should be slightly different than that of an electron because of two particular alternatives: when the electron emits a photon that disintegrates into an electron-positron or muon-antimuon pair, the disintegration creates a pair that is close to or much heavier in mass than the electron. On the other hand, when the muon emits a photon that disintegrates into a muon-antimuon or positron-electron pair, this pair is close to or much lighter in mass than the muon. Experiments confirm this slight difference.*

are still more than eighty decimal places larger than the distances at which the theory alone might run into trouble with infinities (see footnote on p. 129).

We have learned that an electron can couple with a W (see Fig. 85). When a *d*-quark changes into a *u*-quark, emitting a W, can the W then couple with a muon instead of an electron? Yes (see Fig. 90). And what about the anti-neutrino? In the case of the W coupling with a muon, a particle called a mu-neutrino takes the place of the ordinary neutrino (which we will now call an electron neutrino). So now our table of particles has two additional particles next to the electron and the neutrino—the muon and the mu-neutrino.

What about the quarks? Very early on, particles were known that had to be made of heavier quarks than *u* or *d*. Thus a third quark, called *s* (for "strange") was included in the list of fundamental particles. The *s* quark has a mass of about 200 MeV, compared to about 10 MeV for the *u* and *d* quarks.

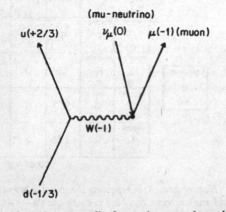

FIGURE 90. *The W has an amplitude to emit a muon instead of an electron. In this case a mu-neutrino takes the place of an electron-neutrino.*

For many years we thought that there were just three "flavors" of quarks—*u*, *d*, and *s*—but in 1974 a new particle called a psi-meson was discovered that could not be made out of the three quarks. There was also a very good theoretical argument that there had to be a fourth quark, coupled to the *s* quark by a W in the same way that the *u* and *d* quark are coupled (see Fig. 91). The "flavor" of this quark is called *c*, and I haven't got the guts to tell you what *c* stands for, but you may have read it in the newspaper. The names are getting worse and worse!

This repetition of particles with the same properties but heavier masses is a complete mystery. What is this strange duplication of the pattern? As Professor I. I. Rabi said of the muon when it was discovered, "Who ordered that?"

Recently another repetition of the list has begun. As we

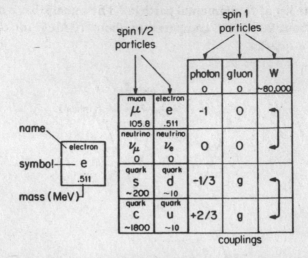

FIGURE 91. *Nature seems to be repeating the spin 1/2 particles. In addition to the muon and mu-neutrino, there are two new quarks—s and c—that have the same charge but higher masses than their counterparts in the next column.*

go to higher and higher energies, Nature seems to keep piling on these particles as if to drug us. I have to tell you about them because I want you to see how apparently complicated the world really looks. It would be very misleading if I were to give you the impression that since we've solved 99% of the phenomena in the world with electrons and photons, that the other 1% of the phenomena will take only 1% as many additional particles! It turns out that to explain that last 1%, we need ten or twenty times as many additional particles.

So here we go again: with even higher energies used in the experiments, an even heavier electron, called the "tau," has been found; it has a mass of about 1,800 MeV, heavy as two protons! A tau-neutrino has also been inferred. And now a funny particle has been found implying a new "flavor" of quark—this time it's "*b*," for "beauty," and it has a charge of − 1/3 (see Fig. 92). Now, I want you to become high-class, fundamental theoretical physicists for a moment, and predict something: a new flavor of quark will be found, called__ (for "____"), with a charge of__, a mass of__ MeV—and we certainly hope it's *true* that it's there![7]

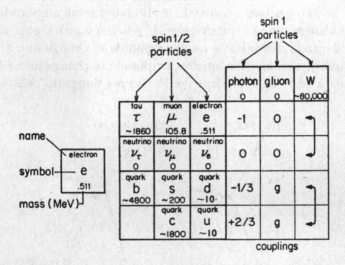

FIGURE 92. *Here we go again! Another repetition of the spin 1/2 particles has begun at even higher energies. This repetition will be complete if a particle with the right properties to imply the existence of a new flavor of quark is found. Meanwhile, preparations are underway to look for the beginning of yet another repetition at even higher energies. What causes these repetitions is a complete mystery.*

Meanwhile, experiments are being done to see if the cycle repeats yet again. At the present time machines are being built to look for an even heavier electron than the tau. If the mass of this supposed particle is 100,000 MeV, they won't be able to produce it. If it is around 40,000 MeV, they might make it.

[7] Since these lectures were given, some evidence has been found for the existence of a *t* quark with a very high mass—around 40,000 MeV.

Mysteries like these repeating cycles make it very interesting to be a theoretical physicist: Nature gives us such wonderful puzzles! Why does She repeat the electron at 206 times and 3,640 times its mass?

I'd like to make one last remark to make things absolutely complete about the particles. When a *d* quark coupling to a W changes into a *u* quark, it also has a small amplitude to change into a *c* quark instead. When a *u* quark goes to a *d* quark, it also has a small amplitude to change into an *s* quark, and an even smaller amplitude to change into a *b* quark (see Fig. 93). Thus the W "screws things up" a little

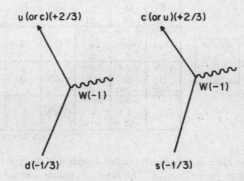

FIGURE 93. *A* d *quark has a small amplitude to change into a* c *quark instead of a* u *quark, and an* s *quark has a small amplitude to change into a* u *quark instead of a* c *quark, with the emission of a W in both cases. Thus the W seems to be able to change a quark's flavor from one column of the table to another (see Fig. 92).*

bit and allows quarks to change from one column of the table to another. Why the quarks have these relative proportions for their amplitude to change to another type of quark is utterly unknown.

So that's everything about the rest of quantum physics. It's a terrible mix-up, and you might say it's a hopeless mess physics has got itself worked into. But it has always looked like this. Nature has always looked like a horrible mess, but as we go along we see patterns and put theories together; a certain clarity comes and things get simpler. The mess I just showed you is much smaller than the mess I would have had to make ten years ago, telling you about the more than four hundred particles. And think about the mess at the beginning of this century, when there was heat, magnetism, electricity, light, X-rays, ultraviolet rays, indices of refraction, coefficients of reflection and other properties of various substances, all of which we have since put together into one theory, quantum electrodynamics.

I would like to emphasize something. The theories about the rest of physics are very similar to the theory of quantum electrodynamics: they all involve the interaction of spin 1/2 objects (like electrons and quarks) with spin 1 objects (like photons, gluons, or W's) within a framework of amplitudes by which the probability of an event is the square of the length of an arrow. Why are all the theories of physics so similar in their structure?

There are a number of possibilities. The first is the limited imagination of physicists: when we see a new phenomenon we try to fit it into the framework we already have—until we have made enough experiments, we don't know that it doesn't work. So when some fool physicist gives a lecture at UCLA in 1983 and says, "This is the way it works, and look how wonderfully similar the theories are," it's not because Nature is *really* similar; it's because the physicists have only been able to think of the same damn thing, over and over again.

Another possibility is that it *is* the same damn thing over and over again—that Nature has only one way of doing things, and She repeats her story from time to time.

A third possibility is that things look similar because they are aspects of the same thing—some larger picture underneath, from which things can be broken into parts that look different, like fingers on the same hand. Many physicists are working very hard trying to put together a grand picture that unifies everything into one super-duper model. It's a delightful game, but at the present time none of the speculators agree with any of the other speculators as to what the grand picture is. I am exaggerating only slightly when I say that most of these speculative theories have no more deep sense to them than your guess about the possibility of a *t* quark, and I guarantee you that they are no better at guessing the mass of a *t* quark than you are!

For example, it appears that the electron, the neutrino, the *d* quark, and the *u* quark all go together—indeed, the first two couple with the W, as do the last two. At present it is thought that a quark can only change "colors" or "flavors." But perhaps a quark could disintegrate into a neutrino by coupling with an undiscovered particle. Nice idea. What would happen? That would mean protons are unstable.

Somebody makes up a theory: The proton is unstable. They make a calculation and find that there would be no protons in the universe anymore! So they fiddle around with the numbers, putting a higher mass into the new particle, and after much effort they predict that the proton will decay at a rate slightly less than the last measured rate the proton has been shown not to decay at.

When a new experiment comes along and measures the proton more carefully, the theories adjust themselves to squeeze out from the pressure. The most recent experiment showed that the proton doesn't decay at a rate that is five times slower than what was predicted in the *last stand* of the theories. What do you think happened? The phoenix just rose again with a new modification of the theory that requires even more accurate experiments to check it. Whether the proton decays or not is not known. To prove that it does not decay is very difficult.

In all of these lectures I did not discuss gravitation. The reason is, gravitational influence between objects is *extremely* small: it is a force that is weaker by 1 followed by 40 zeros than the electrical force between two electrons (perhaps it's 41 zeros). In matter, nearly all of the electrical forces are spent holding the electrons close to the nucleus of their atom, creating a finely balanced mixture of pluses and minuses that cancel out. But with gravitation, the only force is attraction, and it keeps adding and adding as there are more and more atoms until at last, when we get to these ponderously large masses that we are, we can begin to measure the effects of gravity—on planets, on ourselves, and so on.

Because the gravitational force is so much weaker than any of the other interactions, it is impossible at the present time to make any experiment that is sufficiently delicate to measure any effect that requires the precision of a quantum theory of gravitation to explain it.[8] Even though there is no way to test them, there are, nevertheless, quantum theories of gravity that involve "gravitons" (which would appear under a new category of polarizations, called "spin 2"), and other fundamental particles (some with spin 3/2). The best of these theories is not able to include the particles that we do find, and invents a lot of particles that we don't find. The quantum theories of gravity also have infinities in the terms with couplings, but the "dippy process" that is successful in getting rid of the infinities in quantum electrodynamics doesn't get rid of them in gravitation. So not only have we no experiments with which to check a quantum theory of gravitation, we also have no reasonable theory.

[8] When Einstein and others tried to unify gravitation with electrodynamics, both theories were classical approximations. In other words, they were wrong. Neither of these theories had the framework of amplitudes that we have found to be so necessary today.

Throughout this entire story there remains one especially unsatisfactory feature: the observed masses of the particles, m. There is no theory that adequately explains these numbers. We use the numbers in all our theories, but we don't understand them—what they are, or where they come from. I believe that from a fundamental point of view, this is a very interesting and serious problem.

I'm sorry if all this speculation about new particles confused you, but I decided to complete my discussion of the rest of physics to show you how the *character* of those laws—the framework of amplitudes, the diagrams that represent the interactions to be calculated, and so on—appears to be the same as for the theory of quantum electrodynamics, our best example of a good theory.

Note Added in Proofreading, November 1984:

Since these lectures were given, suspicious events observed in experiments make it appear possible that some other particle or phenomenon, new and unexpected (and therefore not mentioned in these lectures), may soon be discovered.

Note Added in Proofreading, April 1985:

At this moment, the "suspicious events" mentioned above appear to be a false alarm. The situation no doubt will have changed again by the time you read this book. Things change faster in physics than in the book publishing business.

Index

lambda particle, 132
lasers, 112
laws: mechanical, 5; of Nature,
   89; Newton's, 5
lens, focusing, 16, 58, 109
light, 13, 23; blue, 33; dim, 14,
   15; in everyday circumstances,
   15; infrared, 13; particles of, 36;
   photon model of, 112; speed of,
   c, 87, 89–90; speed of, in water
   and air, 51; ultraviolet, 13, 149;
   weak monochromatic, 36; white,
   35, 102
line: unit, 62; wavy, 88, 91, 92,
   95, 105
lines, multiplying, 62
lithium metal conducting electricity, 113
location: physical, 105; relative,
   110

m and e, observed values, 127
magic number, 96, 129
magnetic field, 98, 115, 123
magnetic interaction, 131
magnetic moment: of electron, 7,
   115, 118; of neutron, 131; of
   proton, 131, 138
mass(es): calculated (n), 127; of
   heavier particles, 145; of muon,
   143; number, 133; observed
   (m), 151–52; of t quark, 147; of
   tau, 146; of W, 140
material: opaque, 108; reflective, 18
mathematicians, 63
matter, electron theory of, 4
Mautner, Alix, 3
Maxwell, James Clerk, 4; theory, 5
Maya Indians, 11
mesons, 132
MeV, 133
mirage, 52
mirror, 15, 38; etched, 47

monochromatic source, 101–102,
   106
motion, phenomena of, 4
mu-neutrino, 144
muon(s), 143–44; mass of, 143; W
   coupling with a, 144
muon-antimuon pair, 143–44

n, 125
n and j, calculated numbers, 125
Nature: analysis of, 78; laws of,
   89; particle in, 98; phenomenon
   of, 84; strangeness of, 80; variety in, 110
neutral currents, 141, 142
neutral W, 141
neutrino, electron, 144, 145
neutrons, 131
New Zealand, 3
Newton, Sir Isaac, 5, 13–14, 18,
   21–23, 37, 85
nuclear: forces, 131; particles, 9,
   131; phenomena, 8, 77; physics,
   8; reactors, 139
nucleus, 5; atomic, 7; exchanging
   photons, 113
number(s): complex, 63; irrational, 63; junction (j), 91; mass
   (m), 133; m and e, 126–28; mysterious, 126, 130, 135; n and j,
   125–30

oil film, 33, 35
opaque material, 108
optical phenomena, 49

P (A to B), formula for, 88, 90
pair(s): muon-antimuon, 143, 144;
   positron-electron, 116, 119, 126,
   143; quark-antiquark, 139
partial reflection, 15–25, 36, 47,
   64, 66, 69, 72, 75, 77, 100–110;
   colors produced by, 33; depending on thickness of glass,
   22, 34; of many surfaces, 22;

READ MORE IN PENGUIN

In every corner of the world, on every subject under the sun, Penguin represents quality and variety – the very best in publishing today.

For complete information about books available from Penguin – including Puffins, Penguin Classics and Arkana – and how to order them, write to us at the appropriate address below. Please note that for copyright reasons the selection of books varies from country to country.

In the United Kingdom: Please write to *Dept. JC, Penguin Books Ltd, FREEPOST, West Drayton, Middlesex UB7 OBR*

If you have any difficulty in obtaining a title, please send your order with the correct money, plus ten per cent for postage and packaging, to *PO Box No. 11, West Drayton, Middlesex UB7 OBR*

In the United States: Please write to *Penguin USA Inc., 375 Hudson Street, New York, NY 10014*

In Canada: Please write to *Penguin Books Canada Ltd, 10 Alcorn Avenue, Suite 300, Toronto, Ontario M4V 3B2*

In Australia: Please write to *Penguin Books Australia Ltd, 487 Maroondah Highway, Ringwood, Victoria 3134*

In New Zealand: Please write to *Penguin Books (NZ) Ltd,182–190 Wairau Road, Private Bag, Takapuna, Auckland 9*

In India: Please write to *Penguin Books India Pvt Ltd, 706 Eros Apartments, 56 Nehru Place, New Delhi 110 019*

In the Netherlands: Please write to *Penguin Books Netherlands B.V., Keizersgracht 231 NL–1016 DV Amsterdam*

In Germany: Please write to *Penguin Books Deutschland GmbH, Friedrichstrasse 10–12, W–6000 Frankfurt/Main 1*

In Spain: Please write to *Penguin Books S. A., C. San Bernardo 117–6° E–28015 Madrid*

In Italy: Please write to *Penguin Italia s.r.l., Via Felice Casati 20, I–20124 Milano*

In France: Please write to *Penguin France S. A., 17 rue Lejeune, F–31000 Toulouse*

In Japan: Please write to *Penguin Books Japan, Ishikiribashi Building, 2–5–4, Suido, Tokyo 112*

In Greece: Please write to *Penguin Hellas Ltd, Dimocritou 3, GR–106 71 Athens*

In South Africa: Please write to *Longman Penguin Southern Africa (Pty) Ltd, Private Bag X08, Bertsham 2013*

READ MORE IN PENGUIN

PENGUIN SCIENCE AND MATHEMATICS

QED   Richard Feynman
The Strange Theory of Light and Matter

Quantum thermodynamics – or QED for short – is the 'strange theory' – that explains how light and electrons interact. 'Physics Nobelist Feynman simply cannot help being original. In this quirky, fascinating book, he explains to laymen the quantum theory of light – a theory to which he made decisive contributions' – *New Yorker*

God and the New Physics   Paul Davies

Can science, now come of age, offer a surer path to God than religion? This 'very interesting' (*New Scientist*) book suggests it can.

Does God Play Dice?   Ian Stewart
The New Mathematics of Chaos

To cope with the truth of a chaotic world, pioneering mathematicians have developed chaos theory. *Does God Play Dice?* makes accessible the basic principles and many practical applications of one of the most extraordinary – and mindbending – breakthroughs in recent years. 'Engaging, accurate and accessible to the uninitiated' – *Nature*

The Blind Watchmaker   Richard Dawkins

'An enchantingly witty and persuasive neo-Darwinist attack on the anti-evolutionists, pleasurably intelligible to the scientifically illiterate' – Hermione Lee in the *Observer* Books of the Year

The Making of the Atomic Bomb   Richard Rhodes

'Rhodes handles his rich trove of material with the skill of a master novelist ... his portraits of the leading figures are three-dimensional and penetrating ... the sheer momentum of the narrative is breathtaking ... a book to read and to read again' – Walter C. Patterson in the *Guardian*

Asimov's New Guide to Science   Isaac Asimov

A classic work brought up to date – far and away the best one-volume survey of all the physical and biological sciences.

READ MORE IN PENGUIN

PENGUIN SCIENCE AND MATHEMATICS

The Panda's Thumb   Stephen Jay Gould

More reflections on natural history from the author of *Ever Since Darwin*. 'A quirky and provocative exploration of the nature of evolution ... wonderfully entertaining' – *Sunday Telegraph*

Gödel, Escher, Bach: An Eternal Golden Braid   Douglas F. Hofstadter

'Every few decades an unknown author brings out a book of such depth, clarity, range, wit, beauty and originality that it is recognized at once as a major literary event' – Martin Gardner. 'Leaves you feeling you have had a first-class workout in the best mental gymnasium in town' – *New Statesman*

The Double Helix   James D. Watson

Watson's vivid and outspoken account of how he and Crick discovered the structure of DNA (and won themselves a Nobel Prize) – one of the greatest scientific achievements of the century.

The Quantum World   J. C. Polkinghorne

Quantum mechanics has revolutionized our views about the structure of the physical world – yet after more than fifty years it remains controversial. This 'delightful book' (*The Times Educational Supplement*) succeeds superbly in rendering an important and complex debate both clear and fascinating.

Einstein's Universe   Nigel Calder

'A valuable contribution to the demystification of relativity' – *Nature*

Mathematical Circus   Martin Gardner

A mind-bending collection of puzzles and paradoxes, games and diversions from the undisputed master of recreational mathematics.

READ MORE IN PENGUIN

PENGUIN PSYCHOLOGY

Introduction to Jung's Psychology Frieda Fordham

'She has delivered a fair and simple account of the main aspects of my psychological work. I am indebted to her for this admirable piece of work' – C. G. Jung in the Foreword

Child Care and the Growth of Love John Bowlby

His classic 'summary of evidence of the effects upon children of lack of personal attention ... it presents to administrators, social workers, teachers and doctors a reminder of the significance of the family' – *The Times*

The Anatomy of Human Destructiveness Erich Fromm

What makes men kill? How can we explain man's lust for cruelty and destruction? 'If any single book could bring mankind to its senses, this book might qualify for that miracle' – Lewis Mumford

Sanity, Madness and the Family R. D. Laing and A. Esterson

Schizophrenia: fact or fiction? Certainly not fact, according to the authors of this controversial book. Suggesting that some forms of madness may be largely social creations, *Sanity, Madness and the Family* demands to be taken very seriously indeed.

The Social Psychology of Work Michael Argyle

Both popular and scholarly, Michael Argyle's classic account of the social factors influencing our experience of work examines every area of working life – and throws constructive light on potential problems.

Check Your Own I.Q. H. J. Eysenck

The sequel to his controversial bestseller, containing five new standard (omnibus) tests and three specifically designed tests for verbal, numerical and visual–spatial ability.

READ MORE IN PENGUIN

PENGUIN PSYCHOLOGY

Psychoanalysis and Feminism Juliet Mitchell

'Juliet Mitchell has risked accusations of apostasy from her fellow feminists. Her book not only challenges orthodox feminism, however; it defies the conventions of social thought in the English-speaking countries ... a brave and important book' – *New York Review of Books*

Helping Troubled Children Michael Rutter

Written by a leading practitioner and researcher in child psychiatry, a full and clear account of the many problems encountered by young school-age children – development, emotional disorders, underachievement – and how they can be given help.

The Divided Self R. D. Laing

'A study that makes all other works I have read on schizophrenia seem fragmentary ... The author brings, through his vision and perception, that particular touch of genius which causes one to say "Yes, I have always known that, why have I never thought of it before?"' – *Journal of Analytical Psychology*

The Origins of Religion Sigmund Freud

The thirteenth volume in the *Penguin Freud Library* contains Freud's views on the subject of religious belief – including *Totem and Taboo*, regarded by Freud as his best-written work.

The Informed Heart Bruno Bettelheim

Bettelheim draws on his experience in concentration camps to illuminate the dangers inherent in all mass societies in this profound and moving masterpiece.

Introducing Social Psychology Henri Tajfel and Colin Fraser (eds.)

From evolutionary changes to the social influence processes in a given group, a distinguished team of contributors demonstrate how our interaction with others and our views of the social world shape and modify much of what we do.